TensorFlow 机器学习

Machine Learning with TensorFlow

[越] 全华（Quan Hua） [巴基] 沙姆斯·乌尔·阿齐姆（Shams Ul Azeem） 著
[美] 西福·艾哈迈德（Saif Ahmed）

李晗 译

人民邮电出版社

图书在版编目（CIP）数据

TensorFlow机器学习 /（越）全华，（巴基）沙姆斯·乌尔·阿齐姆（Shams Ul Azeem），（美）西福·艾哈迈德（Saif Ahmed）著；李晗 译. -- 北京：人民邮电出版社，2021.5
 ISBN 978-7-115-53125-4

Ⅰ．①T… Ⅱ．①全… ②沙… ③西… ④李… Ⅲ．①人工智能－算法 Ⅳ．①TP18

中国版本图书馆CIP数据核字(2020)第184153号

版权声明

Copyright ©2020 Packt Publishing. First published in the English language under the title Machine Learning with TensorFlow 1.x.
All rights reserved.

本书由英国Packt Publishing公司授权人民邮电出版社出版。未经出版者书面许可，对本书的任何部分不得以任何方式或任何手段复制和传播。

版权所有，侵权必究。

◆ 著　　　[越] 全　华（Quan Hua）
　　　　　[巴基] 沙姆斯·乌尔·阿齐姆（Shams Ul Azeem）
　　　　　[美] 西福·艾哈迈德（Saif Ahmed）
　　译　　　李　晗
　　责任编辑　武晓燕
　　责任印制　王　郁　焦志炜

◆ 人民邮电出版社出版发行　北京市丰台区成寿寺路11号
　邮编　100164　电子邮件　315@ptpress.com.cn
　网址　https://www.ptpress.com.cn
　三河市君旺印务有限公司印刷

◆ 开本：800×1000　1/16
　印张：15.75
　字数：305千字　　　　　　2021年5月第1版
　印数：1－2 000册　　　　　2021年5月河北第1次印刷

著作权合同登记号　图字：01-2017-8641号

定价：79.90元
读者服务热线：(010)81055410　印装质量热线：(010)81055316
反盗版热线：(010)81055315
广告经营许可证：京东市监广登字20170147号

内容提要

TensorFlow 是 Google 所主导的机器学习框架,也是机器学习领域研究和应用的热门对象。

本书主要介绍如何通过 TensorFlow 来构建真实世界的机器学习系统,旨在让读者学以致用,能尽快地上手项目。本书的特色是通过实例来向读者介绍 TensorFlow 的经典知识。本书共有 12 章,包含手写识别器、猫狗分类器、翻译器、文本含义查找、金融中的机器学习、医疗应用等多个实例,完整地向读者展示了实现机器学习应用的全流程。

本书适合想要学习、了解 TensorFlow 和机器学习的读者阅读。如果读者知道基本的机器学习概念,并对 Python 语言有一定的了解,那么能够更加轻松地阅读本书。

作者简介

全华是BodiData（一个身体测量数据平台）的一位计算机视觉和机器学习工程师，专注于为某种手持技术开发计算机视觉和机器学习应用。该技术能够在人体穿着衣服时获取身体虚拟化身。全华获得了越南理工大学的理学学士学位，专攻方向为计算机视觉，并在初创企业中从事计算机视觉和机器学习领域工作。

自2015年起，全华一直在为Packt撰写图书，例如计算机视觉图书 *OpenCV 3 Blueprints*。

我要感谢每一位在本书编写过程中鼓励我的人。

我想向我的合著者、编辑和审稿人表达诚挚的谢意，谢谢他们的建议和帮助。

我要感谢我的家人，特别是我的妻子金思戈（Kim Ngoc）。尽管编写本书占用了大量本该用于陪伴他们的时间，但他们仍然这么支持和鼓励我。是他们让我坚持不懈地编写本书。没有他们的支持，这本书是不可能完成的。

我还要感谢我的老师，是他们传授了我计算机视觉和机器学习领域的知识。

沙姆斯·乌尔·阿齐姆毕业于巴基斯坦国立科技大学电气工程专业。他对计算机科学领域怀有浓厚的兴趣，并以Android开发开启他的技术之旅。目前，他在其他公司从事医疗相关的项目，在机器学习特别是深度学习领域继续追求着自己的职业理想。

他还曾是巴基斯坦国立科技大学RISE实验室的成员。作为 *Designing of Motions for Humanoid Goalkeeper Robots* 的合著者，他在IEEE国际会议ROBIO中享有出版信誉。

西福·艾哈迈德是一位经验丰富的量化分析师，同时也是一位拥有15年行业经验的数据科学家。他的职业生涯始于埃森哲公司的管理咨询，最终在高盛集团和AIG投资公司中担任量化和高级管理职务。最近，他与人共同创立并经营着一家专注于将深度学习应用到医学成像自动化的创业公司。另外，他获得了康奈尔大学的计算机科学学士学位，目前正在加州大学伯克利分校攻读数据科学专业硕士学位。

审稿人简介

纳撒·林茨是一名机器学习研究员,专注于文本分类。在开始接触机器学习时,他主要使用 Theano,但在 TensorFlow 发布后,他就转而使用 TensorFlow。TensorFlow 大大减少了构建机器学习系统的时间,这得益于其直观易懂而强大的神经网络工具集。

我要感谢我的家人和教授,感谢他们给予我的一切帮助。没有他们,我将无法长久保持对软件工程和机器学习的热情。

前言

机器学习彻底改变了现代生活。很多机器学习算法,尤其是深度学习,已经在全球范围内使用,覆盖范围从移动设备到基于云的服务。TensorFlow 是领先的开源软件库之一,能帮助你为各种应用程序构建、训练和部署机器学习系统。这本实操书旨在带给你最好的 TensorFlow,并帮助你构建真实世界的机器学习系统。

读完本书,你会对 TensorFlow 有一个深刻的理解,并能够将机器学习技术应用到你的应用程序中。

本书涵盖内容

第 1 章,初识 TensorFlow,演示如何安装 TensorFlow,并介绍在 Ubuntu、macOS 和 Windows 系统上的安装步骤。

第 2 章,你的第一个分类器,通过实现一个手写识别器,带你享受你的第一次机器学习之旅。

第 3 章,TensorFlow 工具箱,带你浏览 TensorFlow 为实现高效、便捷的工作所提供的工具箱。

第 4 章,猫和狗,教你如何在 TensorFlow 中使用卷积神经网络构建一个图像分类器。

第 5 章,序列到序列模型——你讲法语吗,讨论如何使用序列到序列模型创建一个英语到法语的翻译器。

第 6 章,探索文本含义,通过使用情感分析、实体提取、关键词提取和词关系提取来探索查找文本含义的方法。

第 7 章,利用机器学习赚钱,在本章,你会进入一个充满大量数据的领域:金融世界,本章教你如何使用时间序列数据来解决金融问题。

第 8 章，医疗应用，研究使用深度神经网络解决企业级问题（医疗诊断）的方式。

第 9 章，生产系统自动化，教你如何创建一个生产系统，系统范围从训练模型到服务模型。该系统还可以接收用户反馈，并能够每天自动地进行自我训练。

第 10 章，系统上线，带你遨游亚马逊 Web 服务世界，并展示如何利用亚马逊服务器上的多 GPU 系统。

第 11 章，更进一步——21 个课题，本章介绍了 21 个可以使用深度学习框架 TensorFlow 解决的现实问题。

第 12 章，高级安装，讨论 GPU，并关注 CUDA 的设置步骤及基于 GPU 的 TensorFlow 安装。

阅读本书的前提

在软件方面，整本书都基于 TensorFlow，你可以使用 Linux、Windows 或者 macOS 系统。

在硬件方面，你需要一台可以运行 Ubuntu、macOS 或 Windows 的台式计算机或笔记本计算机。如果你想使用深度神经网络，尤其是当你想要处理大规模数据集时，我们推荐你准备一个 NVIDIA 显卡。

本书的目标读者

如果你已经熟练掌握了机器学习概念、Python 编程、IDE 和命令行，如果你渴望创建一个足够智能和实用的机器学习系统来供现实应用程序使用，那么本书将是你的理想之选。对于那些在工作中熟练掌握编程的人，或者那些需要学习机器学习和 TensorFlow 来支持他们工作的科学家和工程师来说，这本书都将有所帮助。

体例约定

在本书中，你会发现多种用以区分不同信息类型的文本样式。下面是这些样式的一些示例，以及对它们含义的解释。

文本中的代码、数据库表名、文件夹名称、文件名、文件扩展名、路径名、虚拟 URL、用户输入和Twitter的处理是："通过使用include指令，我们可以包含其他上下文信息。"

代码块的样式如下所示。

```
batch_size = 128
  num_steps = 10000
  learning_rate = 0.3
  data_showing_step = 500
```

当我们希望将你的注意力吸引到代码块的某个特定部分时，相关的行或项会设置成粗体。

Layer 1 CONV (32, 28, 28, 4)
Layer 2 CONV (32, 14, 14, 4)
Layer 3 CONV (32, 7, 7, 4)

任何命令行输入或输出都如下所示。

sudo apt-get install python-pip python-dev

新的术语和**重要的词**会用粗体显示。

该图标表示警告或重要说明。

该图标表示提示和技巧。

资源与支持

本书由异步社区出品，社区（https://www.epubit.com/）为您提供相关资源和后续服务。

配套资源

本书提供如下资源：
- 本书源代码；
- 书中彩图文件。

要获得以上配套资源，请在异步社区本书页面中单击 配套资源 ，跳转到下载界面，按提示进行操作即可。注意：为保证购书读者的权益，该操作会给出相关提示，要求输入提取码进行验证。

如果您是教师，希望获得教学配套资源，请在社区本书页面中直接联系本书的责任编辑。

提交勘误

作者和编辑尽最大努力来确保书中内容的准确性，但难免会存在疏漏。欢迎您将发现的问题反馈给我们，帮助我们提升图书的质量。

当您发现错误时，请登录异步社区，按书名搜索，进入本书页面，单击"提交勘误"，输入勘误信息，单击"提交"按钮即可。本书的作者和编辑会对您提交的勘误进行审核，确认并接受后，您将获赠异步社区的 100 积分。积分可用于在异步社区兑换优惠券、样书或奖品。

扫码关注本书

扫描下方二维码，您将会在异步社区微信服务号中看到本书信息及相关的服务提示。

与我们联系

我们的联系邮箱是 contact@epubit.com.cn。

如果您对本书有任何疑问或建议，请您发邮件给我们，并请在邮件标题中注明本书书名，以便我们更高效地做出反馈。

如果您有兴趣出版图书、录制教学视频，或者参与图书翻译、技术审校等工作，可以发邮件给我们；有意出版图书的作者也可以到异步社区在线提交投稿（直接访问 www.epubit.com/selfpublish/submission 即可）。

如果您是学校、培训机构或企业，想批量购买本书或异步社区出版的其他图书，也可以发邮件给我们。

如果您在网上发现有针对异步社区出品图书的各种形式的盗版行为，包括对图书全部或部分内容的非授权传播，请您将怀疑有侵权行为的链接发邮件给我们。您的这一举动是对作者权益的保护，也是我们持续为您提供有价值的内容的动力之源。

关于异步社区和异步图书

"异步社区"是人民邮电出版社旗下IT专业图书社区，致力于出版精品IT技术图书和相关学习产品，为作译者提供优质出版服务。异步社区创办于2015年8月，提供大量精品IT技术图书和电子书，以及高品质技术文章和视频课程。更多详情请访问异步社区官网 https://www.epubit.com。

"异步图书"是由异步社区编辑团队策划出版的精品IT专业图书的品牌，依托于人民邮电出版社近30年的计算机图书出版积累和专业编辑团队，相关图书在封面上印有异步图书的LOGO。异步图书的出版领域包括软件开发、大数据、AI、测试、前端、网络技术等。

异步社区

微信服务号

目录

第1章 初识 TensorFlow ... 1
- 1.1 当前应用 ... 2
- 1.2 安装 TensorFlow ... 2
  - 1.2.1 Ubuntu 安装 ... 2
  - 1.2.2 macOS 安装 ... 4
  - 1.2.3 Windows 安装 ... 5
  - 1.2.4 创建虚拟机 ... 8
  - 1.2.5 测试安装 ... 13
- 1.3 总结 ... 14

第2章 你的第一个分类器 ... 15
- 2.1 关键部分 ... 15
- 2.2 获取训练数据 ... 16
- 2.3 下载训练数据 ... 16
- 2.4 理解分类 ... 17
- 2.5 其他设置 ... 19
- 2.6 逻辑停止点 ... 23
- 2.7 机器学习公文包 ... 23
- 2.8 训练日 ... 27
- 2.9 保存模型以供持续使用 ... 30
- 2.10 为什么隐藏测试集 ... 31
- 2.11 使用分类器 ... 31
- 2.12 深入研究网络 ... 32
- 2.13 所学技能 ... 32
- 2.14 总结 ... 33

第3章 TensorFlow 工具箱 ... 34
- 3.1 快速预览 TensorBoard ... 35
- 3.2 安装 TensorBoard ... 37
  - 3.2.1 嵌入钩子（hook）到代码中 ... 38
  - 3.2.2 AlexNet ... 42
- 3.3 自动化运行 ... 44
- 3.4 总结 ... 45

第4章 猫和狗 ... 46
- 4.1 回顾 notMNIST ... 46
  - 4.1.1 程序配置 ... 47
  - 4.1.2 理解卷积神经网络 ... 48
  - 4.1.3 回顾配置 ... 52
  - 4.1.4 构造卷积神经网络 ... 52
  - 4.1.5 实现 ... 56
- 4.2 训练日 ... 57
- 4.3 真实的猫和狗 ... 59
- 4.4 保存模型以供持续使用 ... 63
- 4.5 使用分类器 ... 64
- 4.6 所学技能 ... 65
- 4.7 总结 ... 65

第5章 序列到序列模型——你讲法语吗 ... 66
- 5.1 快速预览 ... 66

5.2 大量信息 ·· 68
5.3 训练日 ·· 73
5.4 总结 ·· 81

第 6 章 探索文本含义 ································ 82
6.1 额外设置 ·· 83
6.2 所学技能 ·· 96
6.3 总结 ·· 97

第 7 章 利用机器学习赚钱 ···························· 98
7.1 输入和方法 ······································ 98
7.2 处理问题 ·· 101
    7.2.1 下载和修改数据 ·························· 102
    7.2.2 查看数据 ································ 103
    7.2.3 提取特征 ································ 105
    7.2.4 准备训练和测试 ·························· 106
    7.2.5 构建网络 ································ 106
    7.2.6 训练 ···································· 107
    7.2.7 测试 ···································· 108
7.3 更进一步 ·· 108
7.4 个人的实际考虑 ·································· 108
7.5 所学技能 ·· 109
7.6 总结 ·· 110

第 8 章 医疗应用 ···································· 111
8.1 挑战 ·· 112
8.2 数据 ·· 114
8.3 管道 ·· 114
    8.3.1 理解管道 ································ 115
    8.3.2 准备数据集 ······························ 116
    8.3.3 解释数据准备 ···························· 118
    8.3.4 训练流程 ································ 129
    8.3.5 验证流程 ································ 129
    8.3.6 利用 TensorBoard 可视化训练过程 ········ 130
8.4 更进一步 ·· 133
    8.4.1 其他医疗数据挑战 ························ 133
    8.4.2 ISBI 大挑战 ······························ 133
    8.4.3 读取医疗数据 ···························· 134
8.5 所学技能 ·· 138
8.6 总结 ·· 139

第 9 章 生产系统自动化 ······························ 140
9.1 系统概述 ·· 140
9.2 创建项目 ·· 141
9.3 加载预训练模型以加速训练 ························ 142
9.4 为数据集训练模型 ································ 148
    9.4.1 Oxford-IIIT 宠物数据集介绍 ·············· 149
    9.4.2 为训练和测试创建输入管道 ··············· 154
    9.4.3 定义模型 ································ 158
    9.4.4 定义训练操作 ···························· 158
    9.4.5 执行训练过程 ···························· 160
    9.4.6 导出模型以用于生产 ······················ 163
9.5 在生产中利用模型提供服务 ························ 165
    9.5.1 设置 TensorFlow Serving ··············· 166
    9.5.2 运行和测试模型 ·························· 167
    9.5.3 设计 Web 服务器 ······················· 169
9.6 在生产中自动化微调 ······························ 170
    9.6.1 加载用户标记的数据 ······················ 170
    9.6.2 对模型进行微调 ·························· 173
    9.6.3 创建每天运行的 cronjob ················ 179
9.7 总结 ·· 179

第 10 章 系统上线 ·································· 180
10.1 快速浏览亚马逊 Web 服务 ······················ 180
    10.1.1 P2 实例 ································ 181
    10.1.2 G2 实例 ································ 181

10.1.3　F1 实例 ·················· 181
10.1.4　定价 ······················ 182
10.2　应用程序概述 ················ 183
10.2.1　数据集 ···················· 183
10.2.2　准备数据集和输入管道 ······ 184
10.2.3　神经网络架构 ·············· 192
10.2.4　单 GPU 训练流程 ··········· 197
10.2.5　多 GPU 训练流程 ··········· 202
10.3　Mechanical Turk 概览 ········· 209
10.4　总结 ························ 210

第 11 章　更进一步——21 个课题 ··· 211
11.1　数据集和挑战赛 ·············· 211
11.1.1　课题 1：ImageNet 数据集 ··· 211
11.1.2　课题 2：COCO 数据集 ······ 212
11.1.3　课题 3：Open Images 数据集 ························ 212
11.1.4　课题 4：YouTube-8M 数据集 ························ 212
11.1.5　课题 5：AudioSet 数据集 ························ 212
11.1.6　课题 6：LSUN 挑战赛 ······ 213
11.1.7　课题 7：MegaFace 数据集 ························ 213
11.1.8　课题 8：Data Science Bowl 2017 挑战赛 ·············· 213
11.1.9　课题 9：星际争霸游戏 数据集 ···················· 213
11.2　TensorFlow 项目 ·············· 214
11.2.1　课题 10：人体姿态估计 ···· 214
11.2.2　课题 11：对象检测—— YOLO ······················ 214
11.2.3　课题 12：对象检测—— Faster RCNN ·············· 214
11.2.4　课题 13：人体检测—— Tensorbox ················ 214
11.2.5　课题 14：Magenta ········· 215
11.2.6　课题 15：WaveNet ········· 215
11.2.7　课题 16：Deep Speech ····· 215
11.3　有趣的项目 ··················· 215
11.3.1　课题 17：交互式深度着色—— iDeepColor ················ 215
11.3.2　课题 18：Tiny 人脸 检测器 ···················· 215
11.3.3　课题 19：人体搜索 ········ 216
11.3.4　课题 20：人脸识别—— MobileID ················· 216
11.3.5　课题 21：问题回答—— DrQA ······················ 216
11.4　Caffe 转 TensorFlow ··········· 216
11.5　TensorFlow-Slim ·············· 222
11.6　总结 ························ 222

第 12 章　高级安装 ··············· 223
12.1　安装 ························ 223
12.1.1　安装 Nvidia 驱动程序 ······ 224
12.1.2　安装 CUDA 工具箱 ········ 226
12.1.3　安装 cuDNN ·············· 229
12.1.4　安装 TensorFlow ·········· 230
12.1.5　验证支持 GPU 的 TensorFlow ················ 231
12.2　利用 Anaconda 管理 TensorFlow ···················· 231
12.3　总结 ························ 234

第 1 章
初识 TensorFlow

近年来，随着大型公共数据集、廉价图形处理器（Graphics Processing Unit，GPU）的涌现，以及开发者专业程度的增强，机器学习领域已经取得了革命性的成就。作为机器学习的命脉，训练数据的获取和使用也已经变得轻而易举。此外，计算能力的提升也已经能够满足个人甚至小型企业的需求。对于数据科学家来说，当前的十年是激动人心的十年。

行业中使用的一些顶级平台有 Caffe、Theano 和 Torch。虽然基础平台的开发比较活跃，并且是公开共享的，但是由于其安装困难、配置烦琐且难以作为生产解决方案，因此目前其使用者主要局限于机器学习从业者。

在 2015 年底和 2016 年，一些新的平台进入了人们的视野，包括谷歌的 TensorFlow、微软的 CNTK（Computational Network Toolkit）、三星的 Veles 等。由于某些原因，谷歌的 TensorFlow 成为了非常受欢迎的平台。

TensorFlow 是最容易安装的平台之一，它将机器学习技能直接引入了普通爱好者和初级程序员的领域中。与此同时，其高性能特点，如多 GPU 支持，也使其成为了有经验的数据科学家和工业应用领域喜欢的平台。另外，TensorFlow 还提供了一种重构过程和多个用户友好的工具来管理机器学习工作，例如 TensorBoard。最后，该平台还拥有重要的背景支撑，以及来自世界上最大的机器学习基地——谷歌的社区支持。所有这些都是其引人注目的潜在技术优势，后续我们将会深入学习这些技术。

在本章中，我们将学习以下主题。

- macOS X 上 TensorFlow 的安装。
- Windows 和 Linux 上 TensorFlow 的安装，包括核心软件和所有依赖项。
- 开启 Windows 安装的 VM 设置。

1.1 当前应用

尽管 TensorFlow 发布时间不太长，但众多的社区通过努力已经成功地将其移植到现有的机器学习项目中，例如手写识别、语言翻译、动物分类、医疗图像分类，以及情感分析等。机器学习在众多行业及难题中的广泛应用总是能够激起人们的兴趣。有了 TensorFlow，这些问题不仅可以得到解决，而且实现起来也很容易。事实上，在本书中我们将逐步处理并解决前面列举的各个问题。

1.2 安装 TensorFlow

针对多种不同的操作系统，TensorFlow 提供了几种不同且便捷的安装方式和操作方式。基础安装仅仅支持 CPU，而更高级的安装通过将计算推送到单个显卡甚至是多个显卡上，以调动更大的计算力。建议首先从基本的 CPU 安装开始，更复杂的 GPU 和 CUDA 安装将会在第 12 章中讨论。

即使是基础的 CPU 安装，TensorFlow 也提供了多种选项，如下所示。

- 基础的 Python pip 安装。
- 通过 Virtualenv 实现的隔离 Python 安装。
- 通过 Docker 实现的基于容器的完全隔离安装。

推荐使用通过 Virtualenv 实现的隔离 Python 安装，但我们的示例将采用基础的 Python pip 安装，以有助于你关注任务的关键点，即搭建并运行 TensorFlow。

在 Linux 和 macOS 系统中，TensorFlow 可在 Python 2.7 和 Python 3.5 版本下正常工作，而在 Windows 系统中，则只能在 Python 3.5.x 或 Python 3.6.x 版本下使用 TensorFlow。在 Windows 系统中运行一个 Linux 虚拟机，如 Ubuntu 虚拟机，我们也可以在 Python 2.7 环境下使用 TensorFlow。然而，在虚拟机中，无法使用支持 GPU 的 TensorFlow。在 TensorFlow 1.2 版本中，TensorFlow 未提供 macOS 系统下的 GPU 支持。因此，如果你想在 macOS 系统中使用支持 GPU 的 TensorFlow，那么必须从源码进行编译（这超出了本章的讨论范围），否则，仍然可以使用 TensorFlow 1.0 或 TensorFlow 1.1 版本，它们提供了 macOS 下的 GPU 支持。此外，Linux 和 Windows 用户也可以使用支持 CPU 和 GPU 的 TensorFlow。

1.2.1 Ubuntu 安装

Ubuntu 是运行 TensorFlow 最好的 Linux 发行版之一。强烈建议读者使用 Ubuntu 虚拟机，尤其是想要使用 GPU 时。我们将在 Ubuntu 终端上完成大部分的工作。下面，首先通过以下命令来安装 python-pip 和 python-dev。

```
sudo apt-get install python-pip python-dev
```

安装成功界面如图 1-1 所示。

```
ubuntu@ubuntu-PC:~$ sudo apt-get install python-pip python-dev
Reading package lists... Done
Building dependency tree
Reading state information... Done
python-dev is already the newest version (2.7.11-1).
python-pip is already the newest version (8.1.1-2ubuntu0.4).
0 upgraded, 0 newly installed, 0 to remove and 222 not upgraded.
ubuntu@ubuntu-PC:~$
```

图 1-1

如果发现缺失软件包，那么可以通过以下命令进行修正。

```
sudo apt-get update --fix-missing
```

然后，就可以继续安装 python-pip 和 python -dev 了。

现在，准备安装 TensorFlow。我们将安装仅支持 CPU 的 TensorFlow。

通过以下命令启动支持 CPU 的 TensorFlow 的安装。

```
sudo pip install tensorflow
```

安装成功界面如图 1-2 所示。

```
    100% |████████████████████████████████| 890kB 924kB/s
Collecting werkzeug>=0.11.10 (from tensorflow-tensorboard<0.2.0,>=0.1.0->tensor
  Downloading Werkzeug-0.12.2-py2.py3-none-any.whl (312kB)
    100% |████████████████████████████████| 317kB 965kB/s
Collecting setuptools (from protobuf>=3.3.0->tensorflow)
  Downloading setuptools-36.5.0-py2.py3-none-any.whl (478kB)
    100% |████████████████████████████████| 481kB 852kB/s
Installing collected packages: six, funcsigs, pbr, mock, numpy, backports.weakr
5lib, bleach, markdown, setuptools, protobuf, werkzeug, tensorflow-tensorboard,
  Found existing installation: six 1.10.0
    Not uninstalling six at /usr/lib/python2.7/dist-packages, outside environme
  Found existing installation: wheel 0.29.0
    Not uninstalling wheel at /usr/lib/python2.7/dist-packages, outside environ
  Running setup.py install for html5lib ... done
  Running setup.py install for markdown ... done
  Found existing installation: setuptools 20.7.0
    Not uninstalling setuptools at /usr/lib/python2.7/dist-packages, outside en
Successfully installed backports.weakref-1.0.post1 bleach-1.5.0 funcsigs-1.0.2
999 markdown-2.6.9 mock-2.0.0 numpy-1.13.1 pbr-3.1.1 protobuf-3.4.0 setuptools-
.0 tensorflow-1.3.0 tensorflow-tensorboard-0.1.6 werkzeug-0.12.2 wheel-0.30.0
You are using pip version 8.1.1, however version 9.0.1 is available.
You should consider upgrading via the 'pip install --upgrade pip' command.
ubuntu@ubuntu-PC:~$
```

图 1-2

1.2.2 macOS 安装

如果你使用 Python，那么你很可能已经拥有了 Python 包安装器 pip。但是，如果尚未安装 pip，那么可以使用命令 easy_install pip 轻松地安装它。你可能会注意到，我们实际执行的命令是 sudo easy_install pip，之所以需要使用前缀 sudo，是因为此安装需要管理权限。

假设你已经有了基本的软件包安装器 easy_install，如果没有，那么可以从 Python 官网下载并安装它。安装成功将会显示图 1-3 所示的结果。

```
saif — -bash — 80×24
Last login: Thu Feb 18 12:52:14 on ttys001
~ @ alpha-al-ghaib (saif) ::  sudo easy_install pip
Password:
Searching for pip
Best match: pip 8.0.2
Adding pip 8.0.2 to easy-install.pth file
Installing pip script to /Users/saif/anaconda/bin
Installing pip2.7 script to /Users/saif/anaconda/bin
Installing pip2 script to /Users/saif/anaconda/bin

Using /Users/saif/anaconda/lib/python2.7/site-packages
Processing dependencies for pip
Finished processing dependencies for pip
```

图 1-3

接下来，安装 six 包。

```
sudo easy_install --upgrade six
```

安装成功界面如图 1-4 所示。

```
saif — -bash — 80×24
~ @ alpha-al-ghaib (saif) ::  sudo easy_install --upgrade six
Password:
Searching for six
Reading https://pypi.python.org/simple/six/
Best match: six 1.10.0
Processing six-1.10.0-py2.7.egg
six 1.10.0 is already the active version in easy-install.pth

Using /Users/saif/anaconda/lib/python2.7/site-packages/six-1.10.0-py2.7.egg
Processing dependencies for six
Finished processing dependencies for six
```

图 1-4

TensorFlow 的安装仅仅需要满足这两个先决条件，现在可以安装核心平台 TensorFlow 了。此处，将使用前面提到的 `pip` 包安装器，并且可以直接从谷歌网站下载安装 TensorFlow。在本书编写之际，TensorFlow 的最新版本是 v1.3，但是你可以将其修改成想使用的最新版本。

`sudo pip install tensorflow`

pip 安装器将自动收集所有其他所需的依赖项。在该软件安装完全结束之前，你可以看到每个下载项和安装过程信息。

安装成功界面如图 1-5 所示。

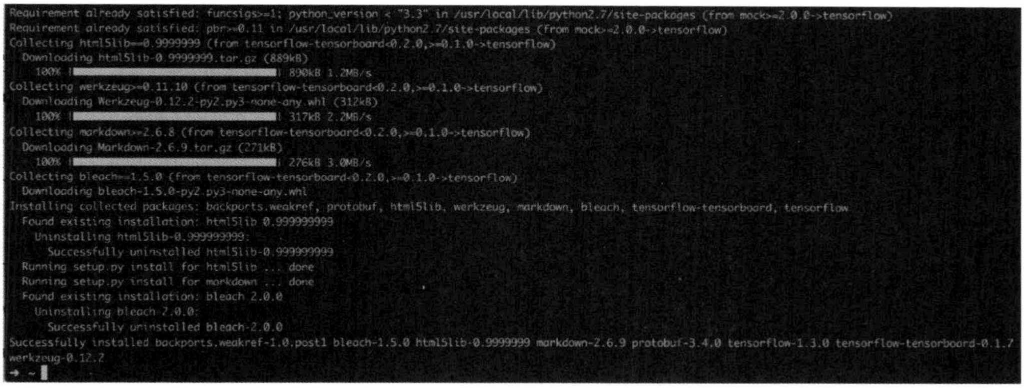

图 1-5

至此，你就可以跳到第 2 章开始训练并运行你的第一个模型了。

如果 macOS X 用户希望完全隔离这些安装，那么可以使用虚拟机来代替，正如在 1.2.3 节中所描述的那样。

1.2.3　Windows 安装

正如前面提到的，在 Windows 系统中，TensorFlow 无法正常工作于 Python 2.7 环境。在本节中，我们将指导你在 Python 3.5 环境下安装 TensorFlow。如果想在 Python 2.7 环境下使用 TensorFlow，需要创建一个 Linux 虚拟机。

首先，需要安装 64 位版本的 Python 3.5.x 或 Python 3.6.x，下载地址为 Python 官网。

请确保下载的是 64 位版本的 Python，其中安装文件的名称中包含 amd64，例如 python-3.6.2-amd64.exe。Python 3.6.2 的安装步骤如图 1-6 所示。

6 第 1 章　初识 TensorFlow

图 1-6

选择 **Add Python 3.6 to PATH** 并单击 **Install Now**。安装过程结束时将会显示图 1-7 所示的画面。

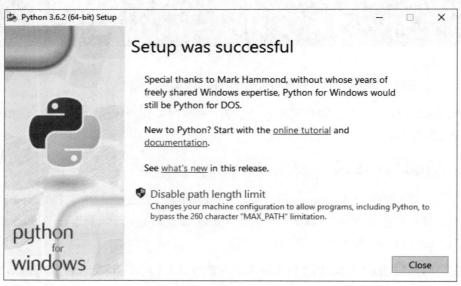

图 1-7

单击 **Disable path length limit**，然后单击 **Close** 完成 Python 的安装。打开 Windows 菜单下的 **Windows PowerShell** 应用，通过下面的命令安装仅支持 CPU 的 TensorFlow 版本。

1.2 安装 TensorFlow

```
pip3 install tensorflow
```

安装过程如图 1-8 所示。

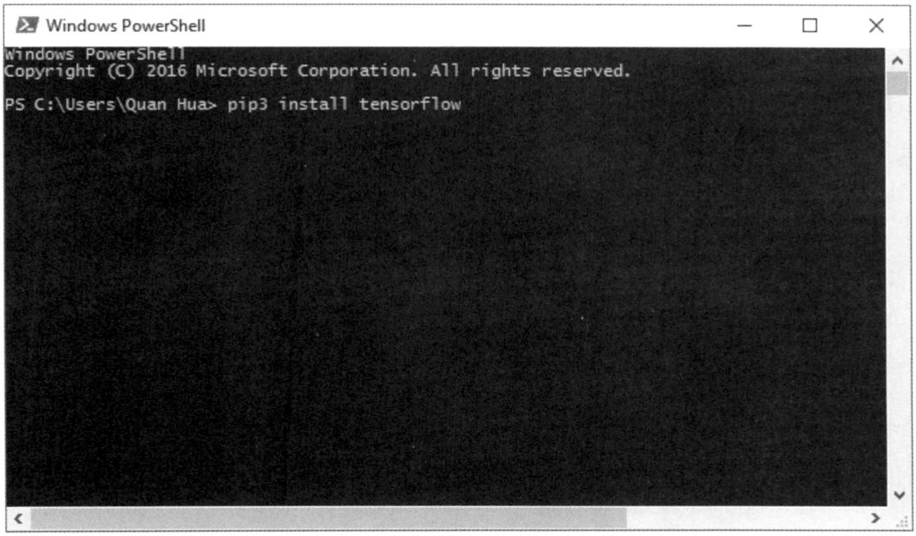

图 1-8

安装结果如图 1-9 所示。

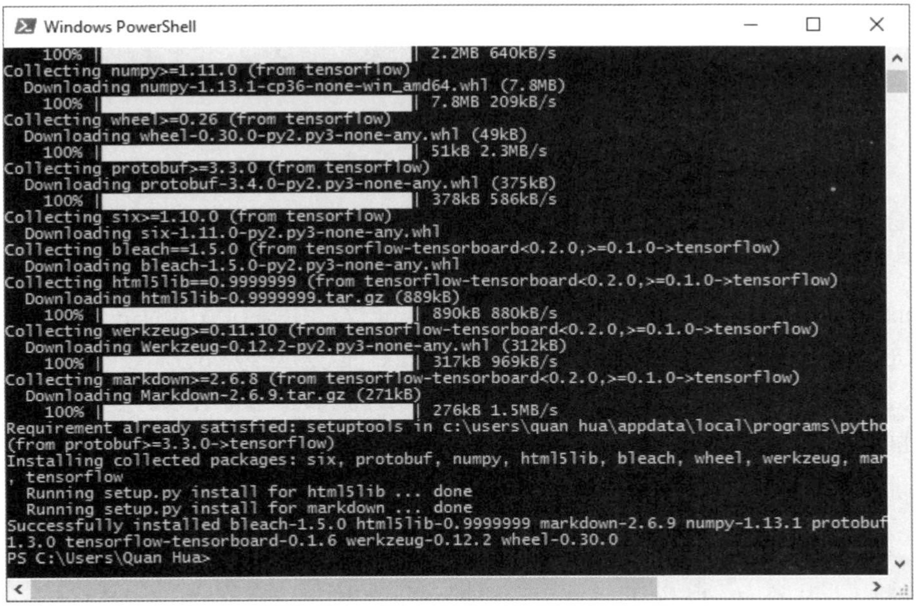

图 1-9

现在你可以在 Windows 系统中，Python 3.5.x 或 3.6.x 环境下使用 TensorFlow 了。

接下来将介绍如何创建一个 Linux 虚拟机，以实现在 Python 2.7 环境下使用 TensorFlow。如果你不需要 Python 2.7，那么可以直接跳到 1.2.5 节。推荐使用免费的 VirtualBox 系统，可在 VirtualBox 官网获得。在本书编写之际（2017 年），VirtualBox 最新的稳定版本是 v5.0.14。成功安装后，你可以运行 **Oracle VM VirtualBox Manager** 控制面板，如图 1-10 所示。

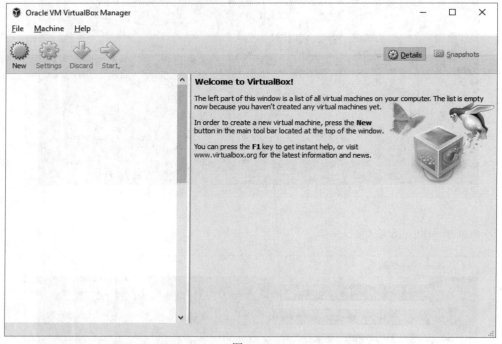

图 1-10

1.2.4　创建虚拟机

　　Linux 有很多不同风格的发行版，但 TensorFlow 文档中提到最多的是 Ubuntu，我们也将使用 Ubuntu Linux。你可以使用任何风格的 Linux，但是应该意识到这一点，即不同风格的 Linux 发行版之间，以及每种发行版的不同版本之间都存在着细微的差别。大多数的差异是良性的，但是有些可能会影响 TensorFlow 的安装，甚至是影响 TensorFlow 的使用。

　　即使在选择了 Ubuntu 之后，也存在很多不同的版本和配置。你可以在 Ubuntu 官网上看到这些信息。

　　我们将安装 Ubuntu 14.04.4 LTS（请确保下载适合你计算机的版本）。标记了 x86 的版本设计为在 32 位机器上运行，而标记了 x64 的版本则设计为运行于 64 位机器上。大多数现代机器都是 64 位的，所以如果你不确定，那就选择后者。

我们利用一个 ISO 文件进行安装，该文件本质上等同于一个安装光盘。Ubuntu 14.04.4 LTS 的 ISO 文件是 ubuntu-gnome-14.04-desktop-amd64.iso。

下载 ISO 安装文件之后，我们将创建一个虚拟机，并使用 ISO 文件在虚拟机上安装 Ubuntu Linux。

在 Oracle VM VirtualBox Manager 上创建虚拟机相当简单，但是需要重点注意其默认选项，因为默认选项并不足以运行 TensorFlow。在安装过程中，将会看到接下来的 8 个界面，最后，它将提示你提供安装文件，该文件正是前面所下载的 ISO 文件。

首先，设置操作系统的类型，并配置分配给虚拟机的随机存储器（RAM）。

1. 注意，我们选择了一个 64 位的安装，因为我们的 Ubuntu 镜像是 64 位的。如果需要，你可以选择使用 32 位镜像，如图 1-11 所示。

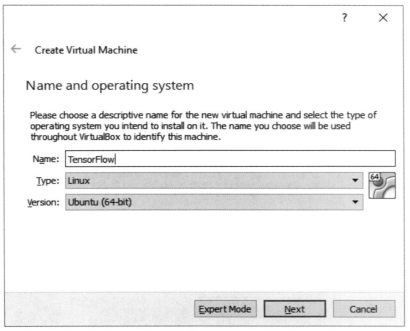

图 1-11

2. 为该虚拟机分配多少内存取决于你的机器有多少内存。在图 1-12 所示的截图中，我们将把一半的 RAM（8GB）分配给虚拟机。需要记住的是，只有运行虚拟机时才会使用这些内存，所以我们可以自由分配内存，我们至少可以分配 4 GB 的内存。

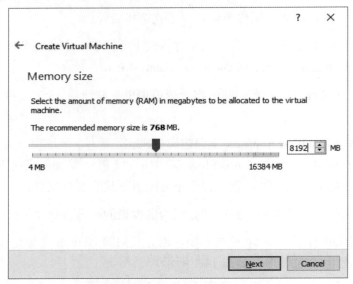

图 1-12

3．虚拟机还需要一个硬盘。我们将创建一个**虚拟硬盘**（**Virtual Hard Disk**，**VHD**），如图 1-13 所示。

图 1-13

4．选择虚拟机的硬盘类型，即**磁盘镜像**（**VirtualBox Disk Image**，**VDI**），如图 1-14 所示。

图 1-14

5．接下来，选择为 VHD（虚拟硬盘）分配多少空间，如图 1-15 所示。这一点很重要，因为我们很快就会使用庞大的数据集。

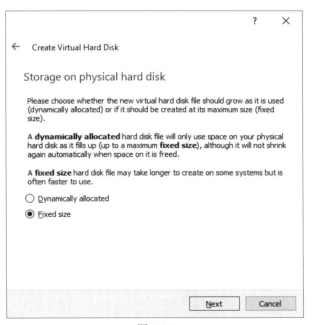

图 1-15

6．TensorFlow 和典型的 TensorFlow 应用程序有一系列的依赖包，例如 NumPy、SciPy 和 pandas。除此以外我们的练习也将下载大型数据集，这些都将用于训练模型。因此，我们为虚拟硬盘分配 12 GB 的空间，如图 1-16 所示。

图 1-16

7．虚拟机设置好之后，它将出现在左边的虚拟机列表中。选中它并单击 **Start** 按钮，如图 1-17 所示。这相当于启动了机器。

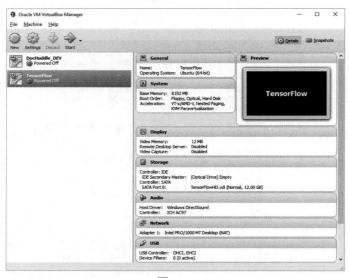

图 1-17

8. 当机器第一次启动时，我们需向其提供安装光盘，即前面下载的 Ubuntu ISO，如图 1-18 所示。

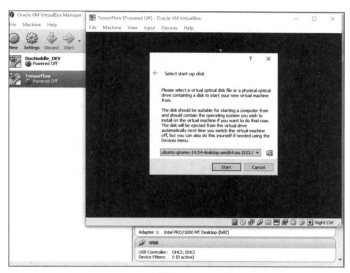

图 1-18

按照安装说明操作，你将拥有一个完整的可供使用的 Ubuntu Linux 虚拟机。在此之后，可以按照 1.2.1 节的内容进行操作。

1.2.5 测试安装

本节将使用 TensorFlow 来执行一个简单的数学运算，以测试安装效果。

首先，在 Linux 或 macOS 系统中打开终端，或者在 Windows 中打开 Windows PowerShell。

然后通过以下命令运行 Python 来使用 TensorFlow。

```
python
```

在 Python shell 中输入以下程序：

```
import tensorflow as tf
a = tf.constant(1.0)
b = tf.constant(2.0)
c = a + b
sess = tf.Session()
print(sess.run(c))
```

运行结果如图 1-19 所示，最后输出计算结果 3.0。

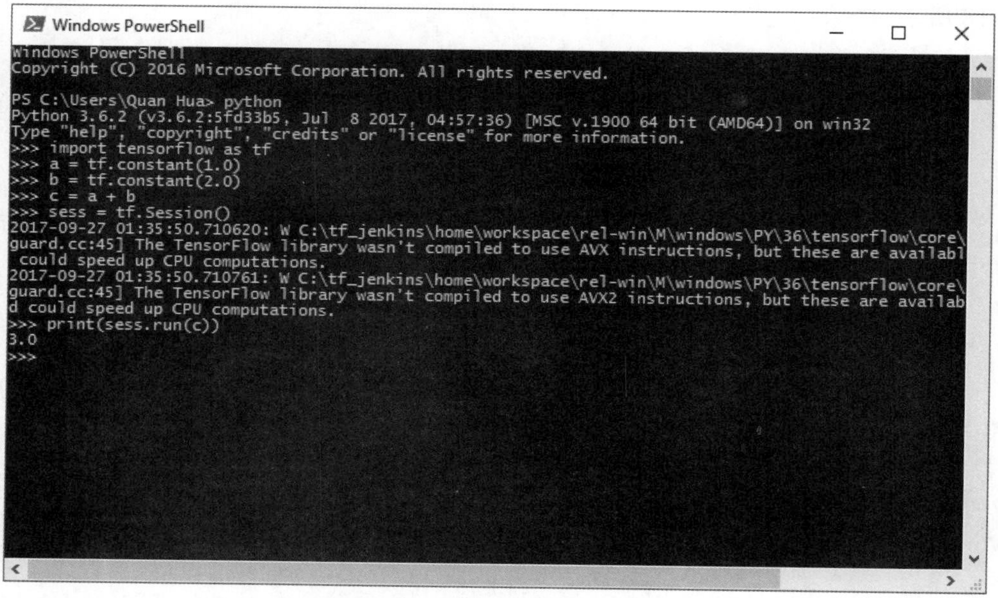

图 1-19

1.3 总结

在本章中，我们介绍了 TensorFlow 在三大主要操作系统上的安装操作步骤。通过这一章的学习，读者应该都已经搭建好并运行了该平台。Windows 用户会面临另一个挑战，这是因为 TensorFlow 在 Windows 系统上只支持 Python 3.5.x 或 Python 3.6.x 的 64 位版本。然而通过这一章的学习，Windows 用户也应该搭建好并运行了 TensorFlow。现在就让我们开始有趣的 TensorFlow 之旅吧！

按照操作步骤安装 TensorFlow 后，下一步是使用一个内置的训练示例来测试安装是否成功。我们将从头开始编写第一个分类器——手写识别器。

在接下来的章节中，我们将回顾 TensorFlow 工具集并在项目中使用它们；还将回顾主要的深度学习概念，并在项目中运用它们。此外，你将有机会尝试多个行业（从金融到医学，再到语言翻译）的项目。

第 2 章 你的第一个分类器

安装好 TensorFlow 后，接下来我们将对其进行检验，为此我们将从头开始编写我们的第一个分类器，并对其进行训练和测试。

我们的第一个分类器是一个手写识别器。在手写识别方面，非常常见的一个训练数据集是 **MNIST** 手写数字数据集。不过，我们将使用一个名为 notMNIST 的类似数据集，它由英语字母表中的前 10 个字母（A～J）的图片组成。

2.1 关键部分

大多数机器学习分类器包含以下 3 个关键部分。

- ◆ 训练管道。
- ◆ 神经网络设置和训练。
- ◆ 应用管道。

训练管道用于获取数据、展示数据、净化数据、均匀化数据，并将数据转换成神经网络可接受的格式。如果训练管道占据你 80%～85%的工作量，那么不要为此感到惊讶，因为这是大多数机器学习工作的实际情况。一般来说，训练数据越真实，训练管道需要花费的时间越多。在企业环境中，训练管道可能是一个持续增强的工作过程，数据集变得更大时更是如此。

第二部分为神经网络设置和训练，对于常规问题它可以快速完成，而对于较困难的问题则需要深入研究。你可能需要一遍又一遍地调整网络设置，直到最终达到所期望的分类器准确率。训练是整个过程中计算成本最高的部分，在评估每次增量修改的结果之前，它都需要花费一定的时间。

一旦完成初始设置，且训练网络达到足够的准确率，那么我们就可以反复使用它了。第 10 章将探索更多高级主题，例如持续学习，在持续学习中对模型的使用可以反馈到分类器的进一步训练中。

2.2 获取训练数据

机器学习离不开训练数据，且通常需要大量的训练数据。机器学习可利用标准训练数据集，这些数据集通常用于基准化节点模型和配置，并提供一种一致性标准来评估相对于前一步的性能变化。此外，很多数据集也用于每年的各种竞赛中。

本章使用由机器学习研究员 Yaroslav Bulatov 提供的训练数据。

2.3 下载训练数据

首先，需要从异步社区中下载训练数据。

后面会利用程序自动下载该数据，但现在我们需要手动下载它，以便查看数据内容并分析数据结构。这一点在设置、编写管道时非常重要，因为我们需要理解数据结构，以便操作数据。

在分析数据时，小数据集是一种理想的选择。可以通过以下命令行或浏览器下载文件，并使用解压缩工具提取目标文件（建议熟悉命令行，因为这些工作都需要实现自动化）。

```
cd ~/workdir
wget http://yaro****upload/notMNIST/notMNIST_small.tar.gz
tar xvf notMNIST_small.tar.gz
```

上面的命令行将会生成一个名为 notMNIST_small 的文件夹，该文件夹包括 10 个子文件夹，每个子文件夹对应字母表中 A~J 的一个字母。在每个对应的字母文件夹中，会有成千上万张像素为 28×28 的字母图片。此外，需要注意的是，每张字母图片的文件名（例如，QnJhbmRpbmcgSXJvbi50dGY=）都是一个不包含使用信息的随机字符串。

2.4 理解分类

我们编写的分类器会试图将未知的图像归为一个类。其中,根据数据内容的不同,可能有以下几种不同的类型。

- 猫或狗。
- 2 或 7。
- 肿瘤或正常。
- 微笑或皱眉。

在本节示例中,我们考虑将每个字母归为一类,所以共有 10 个类。训练集中包含 10 个子文件夹,每个子文件夹下都有成千上万张图片。另外,子文件夹的名称很重要,因为它是其内部每张图片的分类标签。我们的管道将会利用这些细节信息为 TensorFlow 准备数据。

自动化训练数据设置

在理想情况下,我们希望整个过程能够实现自动化。这样在无须随身携带辅助数据的情况下,就可以很容易地在所使用的计算机上端到端地运行整个过程。这一点在后面的工作中十分重要,因为我们经常会在一台计算机上进行开发(开发模型),而在另一台机器上进行部署(生产服务器)。

读者可以在异步社区下载本章所用的代码,以及所有其他章节的代码。为了方便读者理解,我们将在正文中同步编写代码。但是对于比较简单的代码,我们将直接跳过。推荐读者复刻(fork)我们的 GitHub 仓库,并为你的项目复制一份本地副本:

```
cd ~/workdir
git clone https://git****/mlwithtf/MLwithTF
cd chapter_02
```

准备数据集是整个训练过程的一个重要组成部分。在深入分析代码之前,我们将运行 download.py 来自动下载和准备数据集。

```
python download.py
```

运行结果如图 2-1 所示。

```
/notMNIST_small/J
Started loading images from: /home/ubuntu/github/mlwithtf/datasets/notMNIST/test
/notMNIST_small/J
Finished loading data from: /home/ubuntu/github/mlwithtf/datasets/notMNIST/test/
notMNIST_small/J
        Started pickling: J
Finished pickling: J
Finished loading testing data
Started pickling final dataset
Merging train, valid data
Merging test data
('Training set', (200000, 28, 28), (200000,))
('Validation set', (10000, 28, 28), (10000,))
('Test set', (10000, 28, 28), (10000,))
('Compressed pickle size:', 690800514)
Finished pickling final dataset
Finished preparing notMNIST dataset
After reformat:
('Training set', (200000, 784), (200000, 10))
('Validation set', (10000, 784), (10000, 10))
('Test set', (10000, 784), (10000, 10))
(work2) ubuntu@ubuntu-PC:~/github/mlwithtf/chapter_02$
```

图 2-1

现在，查看 download.py 使用的一些函数。

下面的 downloadFile 函数将会自动下载文件，并验证文件大小是否与预期一致。

```
from __future__ import print_function
import os
from six.moves.urllib.request import urlretrieve
import datetime
def downloadFile(fileURL, expected_size):
    timeStampedDir=datetime.datetime.now().strftime("%Y.%m.%d_%I.%M.%S")
    os.makedirs(timeStampedDir)
    fileNameLocal = timeStampedDir + "/" + fileURL.split('/')[-1]
    print ('Attempting to download ' + fileURL)
    print ('File will be stored in ' + fileNameLocal)
    filename, _ = urlretrieve(fileURL, fileNameLocal)
    statinfo = os.stat(filename)
    if statinfo.st_size == expected_size:
        print('Found and verified', filename)
    else:
        raise Exception('Could not get ' + filename)
    return filename
```

可以按如下方式调用此函数。

```
tst_set =
downloadFile('http://yaroslavvb.com/upload/notMNIST/notMNIST_small
.tar.gz', 8458043)
```

提取内容的代码如下所示（注意，需要导入其他包）。

```
import os, sys, tarfile
from os.path import basename

def extractFile(filename):
    timeStampedDir=datetime.datetime.now().strftime("%Y.%m.%d_%I.%M.%S")
    tar = tarfile.open(filename)
    sys.stdout.flush()
    tar.extractall(timeStampedDir)
    tar.close()
    return timeStampedDir + "/" + os.listdir(timeStampedDir)[0]
```

依次调用下载方法和提取方法，如下所示。

```
tst_src='http://yaroslav****/upload/notMNIST/notMNIST_small.tar.gz'
tst_set = downloadFile(tst_src, 8458043)
print ('Test set stored in: ' + tst_set)
tst_files = extractFile(tst_set)
print ('Test file set stored in: ' + tst_files)
```

2.5 其他设置

接下来我们将专注于图像处理和操作。但是这将用到一些额外的库。通过这一点，我们就能理解为何需要安装科学计算所用到的所有经典包了，安装包可以通过以下命令来实现。

```
sudo apt-get install python-numpy python-scipy python-matplotlib
ipython ipython-notebook python-pandas python-sympy python-nose
```

此外，还需要安装图像处理库、一些外部矩阵数学库和需要的其他库，可以通过以下命令来实现。

```
sudo pip install ndimage
sudo apt-get install libatlas3-base-dev gcc gfortran g++
```

将图像转换为矩阵

机器学习的很大一部分工作是矩阵操作。在矩阵操作之前，我们需要将图像转换成一系列的矩阵，即一个宽度等于图像数量的 3 维矩阵。

在本章甚至本书中，几乎所有矩阵运算使用的都是 NumPy 库，它是 Python 中最受欢迎的科学计算包之一。所以，在进行下面一系列操作之前，你应该确保已经安装了 NumPy。

下面的代码打开了图片，并创建了数据矩阵。（注意，此处需要导入额外的 3 个库。）

```python
import numpy as np
from IPython.display import display, Image
from scipy import ndimage

image_size = 28 # Pixel width and height.
pixel_depth = 255.0 # Number of levels per pixel.
def loadClass(folder):
 image_files = os.listdir(folder)
 dataset = np.ndarray(shape=(len(image_files),
image_size,
  image_size), dtype=np.float32)
 image_index = 0
 print(folder)
 for image in os.listdir(folder):
   image_file = os.path.join(folder, image)
   try:
     image_data = (ndimage.imread(image_file).astype(float) -
              pixel_depth / 2) / pixel_depth
     if image_data.shape != (image_size, image_size):
       raise Exception('Unexpected image shape: %s' %
         str(image_data.shape))
     dataset[image_index, :, :] = image_data
     image_index += 1
   except IOError as e: 1
    print('Could not read:', image_file, ':', e, '-
      it\'s ok, skipping.')
   return dataset[0:image_index, :, :]
```

现在，我们获取了上一节提取的文件。接下来，我们会针对所有提取的图像，直接运行以下程序。

```python
classFolders = [os.path.join(tst_files, d) for d in
os.listdir(tst_files) if os.path.isdir(os.path.join(tst_files,
d))]
print (classFolders)
```

```
for cf in classFolders:
    print ("\n\nExaming class folder " + cf)
    dataset=loadClass(cf)
    print (dataset.shape)
```

实际上，该程序只是将字母加载到一个矩阵中，结果如图 2-2 所示。

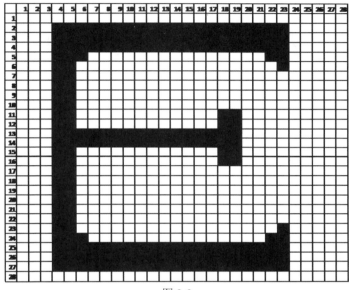

图 2-2

分析该矩阵你会发现很多有趣之处。通过输出堆栈中的任一层（例如，np.set_printoptions(precision=2); print (dataset[47])），就可以得到一个矩阵，如图 2-3 所示。该矩阵中的元素值不是二进制数值，而是浮点数。

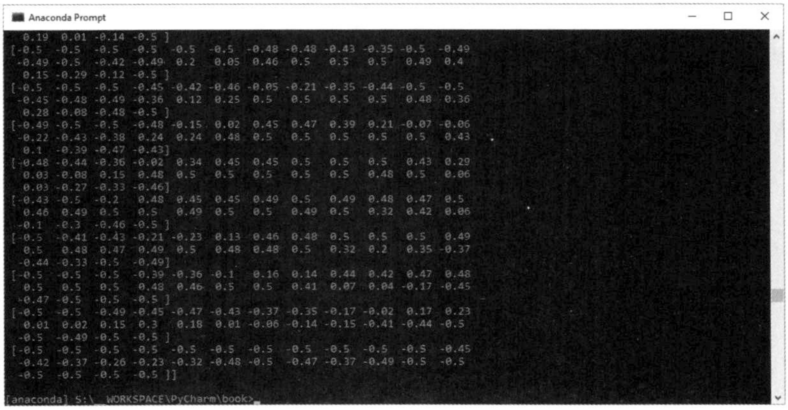

图 2-3

图像首先被加载到一个元素值为 0~255 的矩阵中，如图 2-4 所示。

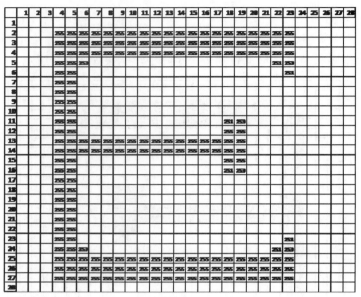

图 2-4

然后，将这些数字缩小至 –0.5~0.5（后文会分析这样做的原因），得到图 2-5 所示的图像。

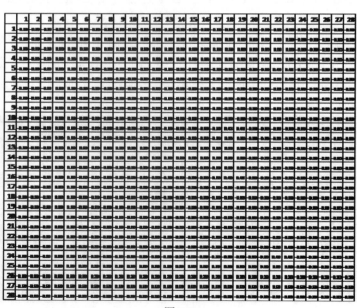

图 2-5

这些都是灰度图像，因此我们只需要处理一层。在后续章节中，我们将处理彩色图像，在那种情况下，每张图片都将对应一个维度为 3 的矩阵，以及一个红色、绿色和蓝色的分离矩阵。

2.6 逻辑停止点

下载训练文件将会花费很长时间，甚至提取所有图片也会花费一定时间。为了避免重复这种费时的操作，我们尝试将所有的工作只做一次，然后创建序列化（pickle）文件。序列化文件是 Python 数据结构的一种归档文件。

下列程序的运行贯穿于训练集和测试集中的每种分类，并为每种分类分别创建一个单独的 pickle 文件。在接下来的学习中，我们先从下列代码开始。

```
def makePickle(imgSrcPath):
    data_folders = [os.path.join(tst_files, d) for d in
     os.listdir(tst_files) if os.path.isdir(os.path.join(tst_files,
     d))]
    dataset_names = []
    for folder in data_folders:
        set_filename = folder + '.pickle'
        dataset_names.append(set_filename)
        print('Pickling %s.' % set_filename)
        dataset = loadClass(folder)
        try:
            with open(set_filename, 'wb') as f:
                pickle.dump(dataset, f, pickle.HIGHEST_PROTOCOL)
        except Exception as e:
            print('Unable to save data to', set_filename, ':', e)
    return dataset_names
```

本质上，pickle 文件是字典的一种可持久化和可重构的转储方式。

2.7 机器学习公文包

我们利用预处理图像创建了简洁、漂亮的 pickle 文件，以用于后续训练和测试我们的分类器。然而，我们最终仅仅得到了 20 个 pickle 文件。此时存在两个问题。首先，有太多易于追踪的文件；其次，这仅仅完成了整个管道的一部分，即只是处理了图像集，但是没有准备 TensorFlow 可以识别的文件。

现在，我们需要创建 3 个主要的数据集：训练集、验证集和测试集。其中，训练集将用来调节分类器，而验证集则用来评估每次迭代的进度，测试集将用来测试我们训练的模型的优劣程度。

完成这些工作的代码量很大，读者可以在 Git 仓库中查看完整的源码，此时需要重点关注以下 3 个函数。

```
def randomize(dataset, labels):
    permutation = np.random.permutation(labels.shape[0])
    shuffled_dataset = dataset[permutation, :, :]
    shuffled_labels = labels[permutation]
    return shuffled_dataset, shuffled_labels

def make_arrays(nb_rows, img_size):
    if nb_rows:
        dataset = np.ndarray((nb_rows, img_size, img_size),
            dtype=np.float32)
        labels = np.ndarray(nb_rows, dtype=np.int32)
    else:
        dataset, labels = None, None
    return dataset, labels

def merge_datasets(pickle_files, train_size, valid_size=0):
 num_classes = len(pickle_files)
 valid_dataset, valid_labels = make_arrays(valid_size,image_size)
 train_dataset, train_labels = make_arrays(train_size,image_size)
 vsize_per_class = valid_size // num_classes
 tsize_per_class = train_size // num_classes

 start_v, start_t = 0, 0
 end_v, end_t = vsize_per_class, tsize_per_class
 end_l = vsize_per_class+tsize_per_class
 for label, pickle_file in enumerate(pickle_files):
   try:
     with open(pickle_file, 'rb') as f:
       letter_set = pickle.load(f)
       np.random.shuffle(letter_set)
       if valid_dataset is not None:
         valid_letter = letter_set[:vsize_per_class, :, :]
         valid_dataset[start_v:end_v, :, :] = valid_letter
         valid_labels[start_v:end_v] = label
         start_v += vsize_per_class
         end_v += vsize_per_class
```

```
            train_letter = letter_set[vsize_per_class:end_l, :, :]
            train_dataset[start_t:end_t, :, :] = train_letter
            train_labels[start_t:end_t] = label
            start_t += tsize_per_class
            end_t += tsize_per_class
    except Exception as e:
        print('Unable to process data from', pickle_file, ':', e)
        raise

    return valid_dataset, valid_labels, train_dataset, train_labels
```

这 3 个函数完成了管道方法。但是我们仍然需要使用管道。先定义训练集、验证集和测试集的大小，你可以改变数据集的大小值，但应该保证它们的大小要小于整个数据集的总大小。

```
train_size = 200000
valid_size = 10000
test_size = 10000
```

这些数据集的大小值将用来构建合并（即组合所有的分类）数据集。我们将以 pickle 文件列表的方式传递数据集的大小，以便导入我们的数据，然后我们会得到一个标签向量和一个图像堆叠的矩阵。最后，将数据集打乱，代码如下所示。

```
valid_dataset, valid_labels, train_dataset, train_labels =
 merge_datasets(
  picklenamesTrn, train_size, valid_size)
_, _, test_dataset, test_labels = merge_datasets(picklenamesTst,
 test_size)
train_dataset, train_labels = randomize(train_dataset,
 train_labels)
test_dataset, test_labels = randomize(test_dataset, test_labels)
valid_dataset, valid_labels = randomize(valid_dataset,
 valid_labels)
```

可以通过下面代码了解新合并的数据集。

```
print('Training:', train_dataset.shape, train_labels.shape)
print('Validation:', valid_dataset.shape, valid_labels.shape)
print('Testing:', test_dataset.shape, test_labels.shape)
```

这是一个工作量巨大的过程，而我们不想再重复这些工作，那么可以再次将 3 个新的数据集序列化到一个单独的、庞大的 pickle 文件中。在后文中，所有的学习都将跳过前面的步骤，并直接利用这个庞大的 pickle 文件。

```
pickle_file = 'notMNIST.pickle'

try:
  f = open(pickle_file, 'wb')
  save = {
    'datTrn': train_dataset,
    'labTrn': train_labels,
    'datVal': valid_dataset,
    'labVal': valid_labels,
    'datTst': test_dataset,
    'labTst': test_labels,
    }
  pickle.dump(save, f, pickle.HIGHEST_PROTOCOL)
  f.close()
except Exception as e:
  print('Unable to save data to', pickle_file, ':', e)
  raise

statinfo = os.stat(pickle_file)
print('Compressed pickle size:', statinfo.st_size)
```

实际上，将矩阵输入到 TensorFlow 的理想方法是将矩阵作为一个一维数组来输入，即将 28×28 的矩阵重新转换成一个包含 784 个数字的字符串。为此，我们将使用下面的 reformat 方法。

```
def reformat(dataset, labels):
  dataset = dataset.reshape((-1, image_size *
    image_size)).astype(np.float32)
  labels = (np.arange(num_labels) ==
    labels[:,None]).astype(np.float32)
  return dataset, labels
```

我们的图像如图 2-6 所示，其中每一行都表示训练集、验证集和测试集中的一张图像。

	1	2	3	4	5	6	7	—	780	781	782	783	784
1	-0.50	-0.50	-0.50	-0.50	-0.50	-0.50	0.50	—	0.50	-0.50	-0.50	-0.50	-0.50
2	-0.50	-0.50	-0.49	-0.50	-0.50	-0.50	0.50	—	0.50	-0.50	-0.50	-0.50	-0.50
3	-0.50	-0.50	-0.49	-0.50	-0.50	-0.50	-0.50	—	0.50	-0.50	-0.50	-0.50	-0.50
—	—	—	—	—	—	—	—	—	—	—	—	—	—
200000	-0.50	-0.50	-0.40	-0.50	-0.50	-0.50	0.50	—	0.50	-0.50	-0.50	-0.50	-0.50
200001	-0.50	-0.50	-0.40	-0.50	-0.50	-0.50	0.50	—	0.50	-0.50	-0.50	-0.50	-0.50
200002	-0.50	-0.50	-0.40	-0.50	-0.50	-0.50	0.50	—	0.50	-0.50	-0.50	-0.50	-0.50
200003	-0.50	-0.50	-0.39	-0.50	-0.50	-0.50	-0.50	—	0.50	-0.50	-0.50	-0.50	-0.50

图 2-6

最后，为了打开并处理 pickle 文件的内容，我们只需读取前面选择的变量名，并像操

作 `hashmap` 一样来提取数据。

```
with open(pickle_file, 'rb') as f:
  pkl = pickle.load(f)
  train_dataset, train_labels = reformat(pkl['datTrn'],
   pkl['labTrn'])
  valid_dataset, valid_labels = reformat(pkl['datVal'],
   pkl['labVal'])
  test_dataset, test_labels = reformat(pkl['datTst'],
   pkl['labTst'])
```

2.8 训练日

训练该模型的完整代码可以在异步社区中获取。

为了训练模型，导入其他几个模块。

```
import sys, os
import tensorflow as tf
import numpy as np
sys.path.append(os.path.realpath('..'))
import data_utils
import logmanager
```

接着，为训练过程定义几个参数。

```
batch_size = 128
num_steps = 10000
learning_rate = 0.3
data_showing_step = 500
```

然后，使用 `data_utils` 包来加载 2.7 节下载的数据集。

```
 dataset, image_size, num_of_classes, num_of_channels =
 data_utils.prepare_not_mnist_dataset(root_dir="..")
 dataset = data_utils.reformat(dataset, image_size,
num_of_channels,
 num_of_classes)
 print('Training set', dataset.train_dataset.shape,
 dataset.train_labels.shape)
 print('Validation set', dataset.valid_dataset.shape,
 dataset.valid_labels.shape)
 print('Test set', dataset.test_dataset.shape,
 dataset.test_labels.shape)
```

我们将从一个全连接网络开始。现在，只需要相信网络设置（稍后会讨论设置的理论依据）。我们将把神经网络描绘成一个计算图，在以下代码中称其为 **graph**。

```
graph = tf.Graph()
with graph.as_default():
 # Input data. For the training data, we use a placeholder that will
 be fed
 # at run time with a training minibatch.
 tf_train_dataset = tf.placeholder(tf.float32,
 shape=(batch_size, image_size * image_size * num_of_channels))
 tf_train_labels = tf.placeholder(tf.float32, shape=(batch_size,
 num_of_classes))
 tf_valid_dataset = tf.constant(dataset.valid_dataset)
 tf_test_dataset = tf.constant(dataset.test_dataset)
 # Variables.
 weights = {
 'fc1': tf.Variable(tf.truncated_normal([image_size * image_size *
 num_of_channels, num_of_classes])),
 'fc2': tf.Variable(tf.truncated_normal([num_of_classes,
 num_of_classes]))
 }
 biases = {
 'fc1': tf.Variable(tf.zeros([num_of_classes])),
 'fc2': tf.Variable(tf.zeros([num_of_classes]))
 }
 # Training computation.
 logits = nn_model(tf_train_dataset, weights, biases)
 loss = tf.reduce_mean(
 tf.nn.softmax_cross_entropy_with_logits(logits=logits,
 labels=tf_train_labels))
 # Optimizer.
 optimizer =
 tf.train.GradientDescentOptimizer(learning_rate).minimize(loss)
 # Predictions for the training, validation, and test data.
 train_prediction = tf.nn.softmax(logits)
 valid_prediction = tf.nn.softmax(nn_model(tf_valid_dataset,
 weights, biases))
 test_prediction = tf.nn.softmax(nn_model(tf_test_dataset, weights,
 biases))
```

The most important line here is the nn_model where the neural network is defined:

```
def nn_model(data, weights, biases):
```

```
layer_fc1 = tf.matmul(data, weights['fc1']) + biases['fc1']
relu_layer = tf.nn.relu(layer_fc1)
return tf.matmul(relu_layer, weights['fc2']) + biases['fc2']
```

在该过程中，用于训练模型的 `loss` 函数也是一个重要因素。

```
loss = tf.reduce_mean(
tf.nn.softmax_cross_entropy_with_logits(logits=logits,
labels=tf_train_labels))
# Optimizer.
optimizer =
tf.train.GradientDescentOptimizer(learning_rate).minimize(loss)
```

`loss` 函数就是与 `learning_rate`(0.3)配合使用的优化器（Stochastic Gradient Descent，SGD，随机梯度下降），同时也是我们试图让其最小化的函数（带有交叉熵的 softmax）。

真正的动作，同时也是最耗时的部分就是最后一个环节——训练循环，如图 2-7 所示。

```
with tf.Session(graph=graph) as session:
    session.run(tf.global_variables_initializer())
    print("Initialized")
    for step in range(num_steps + 1):
        sys.stdout.write('Training on batch %d of %d\r' % (step + 1, num_steps))
        sys.stdout.flush()
        # Pick an offset within the training data, which has been randomized.
        # Note: we could use better randomization across epochs.
        offset = (step * batch_size) % (dataset.train_labels.shape[0] - batch_size)
        # Generate a minibatch.
        batch_data = dataset.train_dataset[offset:(offset + batch_size), :]
        batch_labels = dataset.train_labels[offset:(offset + batch_size), :]
        # Prepare a dictionary telling the session where to feed the minibatch.
        # The key of the dictionary is the placeholder node of the graph to be fed,
        # and the value is the numpy array to feed to it.
        feed_dict = {tf_train_dataset: batch_data, tf_train_labels: batch_labels}
        _, l, predictions = session.run([optimizer, loss, train_prediction], feed_dict=feed_dict)
        if step % data_showing_step == 0:
            acc_minibatch = accuracy(predictions, batch_labels)
            acc_val = accuracy(valid_prediction.eval(), dataset.valid_labels)
            logmanager.logger.info('# %03d  Acc Train: %03.2f%%  Acc Val: %03.2f%% Loss %f' % (
                step, acc_minibatch, acc_val, l))
    logmanager.logger.info("Test accuracy: %.1f%%" % accuracy(test_prediction.eval(), dataset.test_labels))
```

图 2-7

在 chapter_02 目录中，可以利用下面的命令来运行训练进程。

python training.py

运行该进程将产生图 2-8 所示的输出。

```
Initialized
2017-09-29 03:33:28,900 - MLwithTF - INFO - # 000  Acc Train: 11.72%  Acc Val: 18.66% Loss 33.751575
2017-09-29 03:33:29,295 - MLwithTF - INFO - # 500  Acc Train: 68.75%  Acc Val: 69.35% Loss 1.133016
2017-09-29 03:33:29,683 - MLwithTF - INFO - # 1000  Acc Train: 66.41%  Acc Val: 68.23% Loss 0.997880
2017-09-29 03:33:30,101 - MLwithTF - INFO - # 1500  Acc Train: 77.34%  Acc Val: 75.75% Loss 0.744131
2017-09-29 03:33:30,491 - MLwithTF - INFO - # 2000  Acc Train: 69.53%  Acc Val: 76.69% Loss 1.102222
2017-09-29 03:33:30,897 - MLwithTF - INFO - # 2500  Acc Train: 69.53%  Acc Val: 77.12% Loss 0.942900
2017-09-29 03:33:31,290 - MLwithTF - INFO - # 3000  Acc Train: 80.47%  Acc Val: 77.96% Loss 0.652119
2017-09-29 03:33:31,689 - MLwithTF - INFO - # 3500  Acc Train: 77.34%  Acc Val: 78.18% Loss 0.738713
2017-09-29 03:33:32,088 - MLwithTF - INFO - # 4000  Acc Train: 75.78%  Acc Val: 75.89% Loss 0.811817
2017-09-29 03:33:32,486 - MLwithTF - INFO - # 4500  Acc Train: 76.56%  Acc Val: 71.15% Loss 0.767980
2017-09-29 03:33:32,890 - MLwithTF - INFO - # 5000  Acc Train: 78.91%  Acc Val: 79.68% Loss 0.725805
2017-09-29 03:33:33,291 - MLwithTF - INFO - # 5500  Acc Train: 78.91%  Acc Val: 79.76% Loss 0.720853
2017-09-29 03:33:33,688 - MLwithTF - INFO - # 6000  Acc Train: 81.25%  Acc Val: 79.60% Loss 0.630755
2017-09-29 03:33:34,094 - MLwithTF - INFO - # 6500  Acc Train: 81.25%  Acc Val: 79.94% Loss 0.717576
2017-09-29 03:33:34,483 - MLwithTF - INFO - # 7000  Acc Train: 82.03%  Acc Val: 79.53% Loss 0.730982
2017-09-29 03:33:34,870 - MLwithTF - INFO - # 7500  Acc Train: 82.81%  Acc Val: 80.48% Loss 0.606189
2017-09-29 03:33:35,257 - MLwithTF - INFO - # 8000  Acc Train: 78.91%  Acc Val: 79.64% Loss 0.683653
2017-09-29 03:33:35,647 - MLwithTF - INFO - # 8500  Acc Train: 82.81%  Acc Val: 80.42% Loss 0.643934
2017-09-29 03:33:36,023 - MLwithTF - INFO - # 9000  Acc Train: 83.59%  Acc Val: 80.34% Loss 0.608325
2017-09-29 03:33:36,420 - MLwithTF - INFO - # 9500  Acc Train: 82.81%  Acc Val: 80.38% Loss 0.595343
2017-09-29 03:33:36,820 - MLwithTF - INFO - # 10000  Acc Train: 74.22%  Acc Val: 80.62% Loss 0.943106
2017-09-29 03:33:36,840 - MLwithTF - INFO - Test accuracy: 87.4%
(work2) ubuntu@ubuntu-PC:~/github/mlwithtf/chapter_02$
```

图 2-8

我们运行了很多次循环，并每隔 500 次循环输出指示性的结果。你也可以修改这些设置中的任何参数。其中，用户需要重点关注循环。

- 多次循环该过程。
- 每一次循环都会创建一小批图像，它们是完整图像集的一部分。
- 每一步都会运行 TensorFlow 会话，并产生一个损失值和一组预测值，还会对验证集进行预测。
- 在迭代循环的末尾，将对测试集做出最终的预测。
- 对于所做的每一个预测，将以预测准确率的形式来观察进展。

之前并没有讨论过 accuracy 方法，该方法通过简单地对比预测标签和已知标签来计算准确率。

```
def accuracy(predictions, labels):
 return (100.0 * np.sum(np.argmax(predictions, 1) ==
   np.argmax(labels, 1))
         / predictions.shape[0])
```

运行前面的分类器将会产生 85%左右的准确率，这个结果是很不错的，毕竟才刚刚开始，我们还可以继续做更多的调整。

2.9　保存模型以供持续使用

为了保存 TensorFlow 会话中的变量以供将来使用，我们可以使用 saver() 函数，如下所示。

```
saver = tf.train.Saver()
```

之后,可以通过恢复下面的检查点来恢复模型的状态,以避免乏味的重复训练。

```
ckpt = tf.train.get_checkpoint_state(FLAGS.checkpoint_dir)
if ckpt and ckpt.model_checkpoint_path:
saver.restore(sess, ckpt.model_checkpoint_path)
```

2.10 为什么隐藏测试集

需要注意的是,在进行最后一步之前,我们都没有使用测试集,为什么不使用呢?这是一个很重要的细节,该细节能确保测试数据集仍旧是一个良好的测试数据集。当迭代训练集并以某种方式训练分类器时,我们可能会将分类器运用于每张图像,或者过度训练。当你学习训练集而非学习每种类别包含的特征时,通常会出现这种情况。

当过度训练时,在训练集上迭代过程的准确性看起来将会很不错,但这都是虚假预期。引入一个从未见过的测试集则会将真实情况带入到该过程中。如果训练集的准确率很高,而测试集的准确率很差,这就意味着出现了过度拟合的问题。

这就是我们保留了一个独立测试集的原因。它有助于反映出分类器的真实准确率。这也是你不应该将数据集混合或者将数据集与测试集混在一起的原因。

2.11 使用分类器

我们将通过 notMNIST_small.tar.gz 来说明分类器的用法,该数据集可作为分类器的测试集。为了持续使用分类器,你可以使用自己的图像,并通过类似的管道进行测试而非训练。

你可以自己创建一些 28 × 28 的图像,并将它们放入测试集进行评估。

实际使用中出现的问题是自然环境下图像的不均匀性。这可能需要你找到图像,然后裁剪它们、缩小它们,或者执行一系列其他转换。这一切都归属于之前讨论过的管道使用。

另一种涉及较大图像的处理技术(例如在页面大小的图像上查找一个字母)就是在大图像上用一个小窗口滑动,并将图像的每个子块输入分类器。

在后续章节中,我们将会把模型引入生产环境中。在实际引入之前,常见的预览设置是将训练后的模型迁移到云服务器中。系统的前台可能是一款智能手机应用程序,它可以拍摄照片并将其发送到后台进行分类。在这种情况下,我们将使用 Web 服务来包装整个程序，以接受输入的分类请求，并自动响应。流行的设置有很多，我们将在第 9 章讲解其中的几种。

2.12 深入研究网络

准确率 86%，这对于仅仅花费两小时的工作来说，是一个很厉害的结果，但是我们可以做得更好。未来的重点很大程度上在于改进神经网络。前面的应用程序使用了一个全连接设置，其中网络层上的每个节点都与前一层的所有节点相连接，如图 2-9 所示。

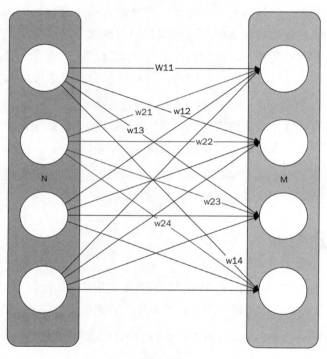

图 2-9

在接下来的章节中，我们将学习更加复杂的网络设置，复杂的网络设置很快捷但不够理想。最大的问题是它拥有大量的参数，这可能导致模型在训练数据上过度拟合。

2.13 所学技能

在本章中，你应该学到了这些技能。

- 准备训练和测试数据。
- 创建一个 TensorFlow 可使用的训练集。
- 建立一个基础的神经网络图。
- 训练 TensorFlow 分类器。
- 验证分类器。
- 向管道输入真实数据。

2.14 总结

在本章中,我们创建了一个手写分类器。同时,还创建了一个完整的管道来实现训练设置和执行的完全自动化。这意味着我们的程序可以移植到几乎任何服务器上,并完全可能像一整套系统一样继续运行。

第 3 章 TensorFlow 工具箱

有些机器学习平台面向的是学术或工业领域的科学家和从业者。因此，它们虽然相当强大，但通常是有瑕疵的，并且几乎没有什么用户体验。

在不同阶段观察模型，以及通过模型分析和运行查看并聚合性能，需要花费相当多的精力。即使是查看神经网络也要付出比预期更多的精力。

当神经网络比较简单且深度只有几层时我们还可以勉强接受，但现在的网络要复杂得多。在 2015 年，微软使用一个深度为 152 层的网络赢得了每年一度的 ImageNet 大赛。可视化这样的网络可能会非常困难，而查看权重和偏差的困难更为巨大。

从业者开始使用自制的可视化工具和引导工具来分析网络和运行性能。TensorFlow 在发布整体平台时直接发布了 TensorBoard，从而改变了这一状况。TensorBoard 无须任何额外的安装和设置即可运行。

用户只需根据他们想要捕捉的内容来编写代码。TensorBoard 的特性包括绘制随时间变化的事件、学习率和损失值，权重和偏差的直方图以及图像。图形浏览器（graph explorer）允许对神经网络进行交互式检查。

本章将关注以下几个方面。

- ◆ 以 4 种常见的模型和数据集为例，我们将介绍使用 TensorBoard 所用的指令，并突出显示需要更改的内容。
- ◆ 然后，回顾捕获的数据，并解析它的方式。
- ◆ 最后，回顾图形浏览器所显示的普通图形。这将有助于将常见的神经网络设置可视化，这一点会在后续章节中介绍，这也是对常见网络的可视化介绍。

3.1 快速预览 TensorBoard

即使没有安装 TensorFlow，你也可以使用 TensorBoard 的参考实现。

先下载 CIFAR-10 数据集。CIFAR-10 数据集由 Alex Krizhevsky、Vinod Nair 和 Geoffrey Hinton 编译，包含 10 种类别的 60 000 张图片。该数据集已经成为机器学习领域的标准学习工具和基准数据集之一。

我们从图形浏览器的学习开始，使用卷积网络对图像进行分类，如图 3-1 所示。

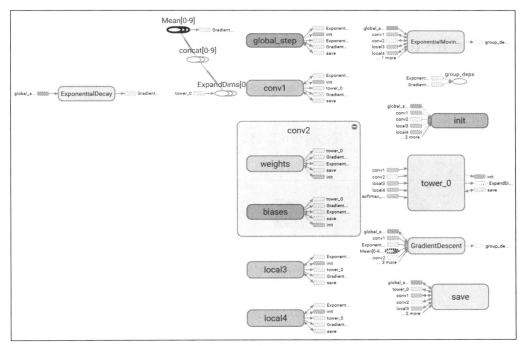

图 3-1

这仅仅是一种可能的图形视图。你也可以尝试图形浏览器，它允许我们深入分析各个组件。

EVENTS 选项卡也可用于快速预览。该选项卡显示了标量数据随时间变化的信息，不同的统计数据会被分组到右边的各个选项卡中。图 3-2 所示的截图显示了一些通用的标量统计信息，例如损失值、学习率、交叉熵，以及网络中多个部分的稀疏性等。

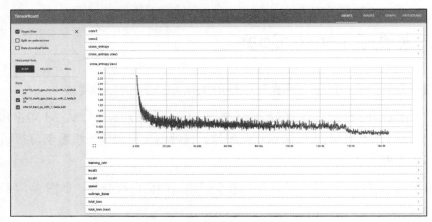

图 3-2

接下来是 **HISTOGRAMS** 选项卡。与 EVENTS 选项卡类似，它显示了张量数据随时间变化的信息。抛开其名称，正如 TensorFlow v0.7 中所介绍的，它实际上并没有显示直方图，相反，它以百分位数的格式显示了张量数据的摘要信息。

图 3-3 显示了摘要视图。与 EVENTS 选项卡类似，数据也被分组到右侧的选项卡中。我们可以开启或关闭不同的运行状态，也可以将运行状态以平铺的方式来显示，以此实现有趣的比较。

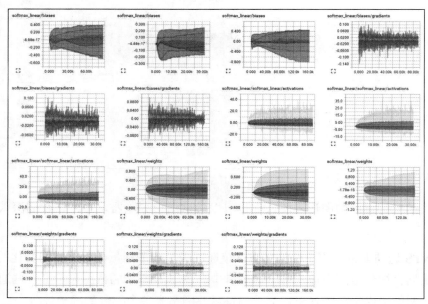

图 3-3

我们可以在图 3-2 所示的界面左侧看到 TensorBoard 描述了 3 个运行状态，但是此处我们仅仅关注 softmax 函数及其相关参数。

现在只需查看我们能对分类器做些什么。

虽然，摘要视图并不能充分发挥 **HISTOGRAMS** 选项卡的价值。所以，我们将放大一个图表来观察发生了什么，如图 3-4 所示。

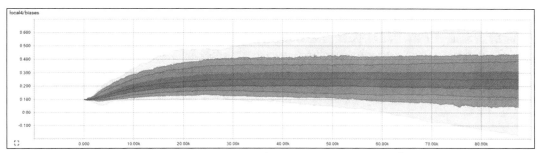

图 3-4

从图 3-4 可以注意到，每个柱状图都显示了由 9 条线组成的时间序列。其中，顶部是最大值，中部是中间值，底部是最小值，中间值以及紧挨着中间的 2 条直线（上、下各一条）分别表示 $\frac{3}{2}$ 标准差、1 标准差和 $\frac{1}{2}$ 标准差标记。

显然，这表示的是多重模态分布，因为它并不是一个直方图。然而，它确实提供了一个快速的方法，让我们可以避免海量数据的转换。

需要注意的一些事项是：运行状态是如何收集和隔离数据的，如何收集不同的数据流，如何扩大视图，以及如何放大每个图表。

有了足够的图形后，下面我们将深入到代码中，以实现自动运行。

3.2 安装 TensorBoard

TensorFlow 预先包含了 TensorBoard，从而 TensorBoard 已经安装完成。它作为一个本地服务的 Web 应用而运行，并且可以通过在浏览器中输入 http://0.0.0.0:6006 来访问。此外，TensorBoard 不需要服务端代码或其他配置。

根据路径的位置，你可以直接运行它，如下所示。

```
tensorboard --logdir=/tmp/tensorlogs
```

如果你的路径不正确，那么可能需要对应用程序添加相应的前缀，如下面命令行所示。

```
tf_install_dir/ tensorflow/tensorboard --
logdir=/tmp/tensorlogs
```

在 Linux 上，可以在后台运行 TensorBoard，并让它持续运行，如下所示。

```
nohup tensorboard --logdir=/tmp/tensorlogs &
```

不过，有些想法应该体现在目录结构中。界面左侧的 **Runs** 列表由 logdir 位置的子目录驱动。图 3-5 显示了两个运行状态：MNIST_Run1 和 MNIST_Run2。一个有组织的 runs 文件夹能够并排画出连续的运行状态，以观察它们之间的不同之处。

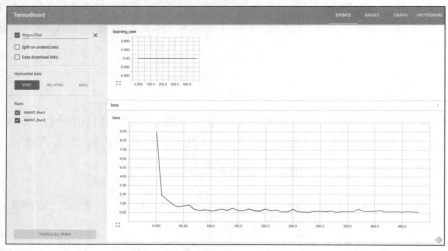

图 3-5

初始化 `writer` 时，需要将日志目录作为第一个参数输入函数中，代码如下所示。

```
writer = tf.summary.FileWriter("/tmp/tensorlogs",
sess.graph)
```

此时我们需要考虑保存一个基础的文件夹，并为每个运行状态添加其特定的子目录，这将有助于组织输出，而无须进行更多的思考。后续我们将对其进行更多讨论。

3.2.1 嵌入钩子（hook）到代码中

开始使用 TensorBoard 的最好方法是使用现有的工作示例，并向其中添加使用 TensorBoard 所需要的代码。后文将会在几个常用的训练脚本中实现这一操作。

下面，我们从机器学习的"Hello World"——MNIST 手写数字分类练习开始。

3.2 安装 TensorBoard

 目前正在使用的 MNIST 数据库包含 60 000 张用于训练的图片，以及 10 000 张用于测试的图片。它最初是由 Chris Burges 和 Corinna Cortes 收集的，并由 Yann LeCun 扩充。可以在 Yann LeCun 的网站上找到更多关于此数据库的信息。

TensorFlow 附带了一个测试脚本，它使用 MNIST 手写数字数据集演示了一个卷积神经网络。

我们修改该脚本，使其支持 TensorBoard。该脚本可在异步社区官网下载。

为了理解整个过程，建议你跟随本节逐步地进行修改。

首先，在 `main` 类的前面部分，定义 `convn_weights`、`convn_biases` 等参数。紧接着，我们将编写下面的代码，以将它们添加到 `histogram` 中。

```
tf.summary.histogram('conv1_weights', conv1_weights)
tf.summary.histogram('conv1_biases', conv1_biases)
tf.summary.histogram('conv2_weights', conv2_weights)
tf.summary.histogram('conv2_biases', conv2_biases)
tf.summary.histogram('fc1_weights', fc1_weights)
tf.summary.histogram('fc1_biases', fc1_biases)
tf.summary.histogram('fc2_weights', fc2_weights)
tf.summary.histogram('fc2_biases', fc2_biases)
```

前面几行代码为 **HISTOGRAMS** 选项卡捕获了相应的值。这里需要注意 **HISTOGRAMS** 选项卡子栏目中的已捕获的值，图 3-6 展示了这些值。

接下来，我们记录一些 `loss` 数据。

```
loss += 5e-4 * regularizers
```

在上一行代码后，为 `loss` 数据添加 `scalar` 摘要。

```
tf.summary.scalar("loss", loss)
```

计算学习率 `learning_rate`。

```
learning_rate = tf.train.exponential_decay(
    0.01,  # Base learning rate.
    batch * BATCH_SIZE,  # Current index into the
    dataset.
    train_size,  # Decay step.
    0.95,  # Decay rate.
    staircase=True)
```

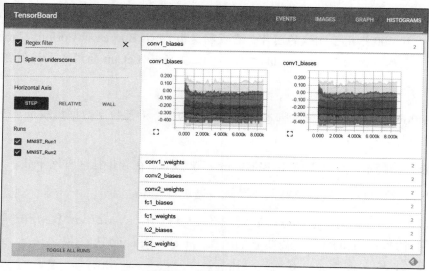

图 3-6

为 learning_rate 数据添加 scalar 摘要。

tf.summary.scalar("learning_rate", learning_rate)

上述代码将数据捕获到 **EVENTS** 选项卡中的标量矩阵中,如图 3-7 所示。

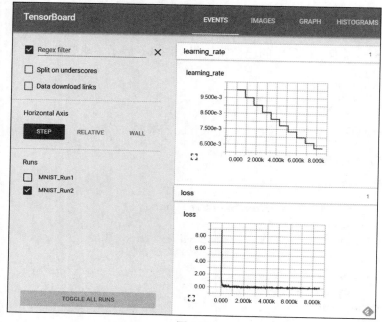

图 3-7

修改脚本来保存图形设置。找到脚本中创建会话的代码部分。

```
# Create a local session to run the training.
start_time = time.time()
with tf.Session() as sess:
```

定义 `sess` 句柄之后,我们将使用以下代码捕获图形。

```
writer = tf.summary.FileWriter("/tmp/tensorlogs", sess.graph)
merged = tf.summary.merge_all()
```

运行会话时需要添加 `merged` 对象,代码如下所示。

```
l, lr, predictions = sess.run([loss, learning_rate,
train_prediction], feed_dict=feed_dict)
# Run the graph and fetch some of the nodes.
sum_string, l, lr, predictions = sess.run([merged,
loss,
learning_rate, train_prediction],
feed_dict=feed_dict)
```

最后,需要在指定的步骤中编写摘要,类似于定期输出验证集准确率等。在计算 `sum_string` 之后,需要再添加一行代码。

```
writer.add_summary(sum_string, step)
```

通过以上代码我们捕获了损失值和学习率,以及神经网络中的关键中间参数和计算图结构。到此,我们已经查看了 **EVENTS** 和 **HISTOGRAMS** 选项卡,下面看一下 **GRAPH** 选项卡,如图 3-8 所示。

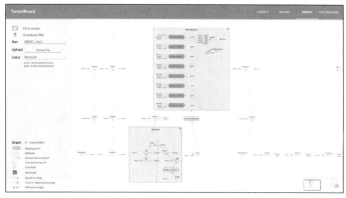

图 3-8

3.2.2 AlexNet

接触过深度学习处理图像的人应该熟悉 AlexNet。具有里程碑意义的论文"ImageNet Classification with Deep Convolutional Neural Networks"引入了 AlexNet 网络概念，该论文由 Alex Krizhevsky、Ilya Sutskever 和 Geoffrey E. Hinton 发表。

AlexNet 网络架构（见图 3-9）在年度 ImageNet 竞赛上取得了相当高的准确率。我们将在后续章节中使用该网络架构，现在使用 TensorBoard 来浏览该网络架构。

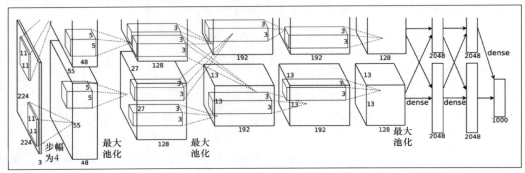

图 3-9

我们不会逐行讲解现有 AlexNet 的代码的修改，读者可以通过对比谷歌提供的原始模型代码与本书代码库中包含的修改后的代码来查看二者之间的差异。

对比之后可以看出，包含 TensorBoard 的代码所引入的更改类似于我们在 MNIST 示例中所做的修改。

首先，找到这段代码的位置。

```
sess = tf.Session(config=config)
sess.run(init)
```

然后，用下面的代码替换它。

```
sess = tf.Session(config=config)
writer = tf.summary.FileWriter("/tmp/alexnet_logs", sess.graph)
sess.run(init)
```

最后，运行 Python 文件 alexnet_benchmark.py 和 TensorBoard 命令来可视化图形。

```
python alexnet_benchmark.py
tensorboard --logdir /tmp/alexnet_logs
```

这部分我们仅仅关注计算图，图 3-10 显示了图形浏览器的一部分内容。

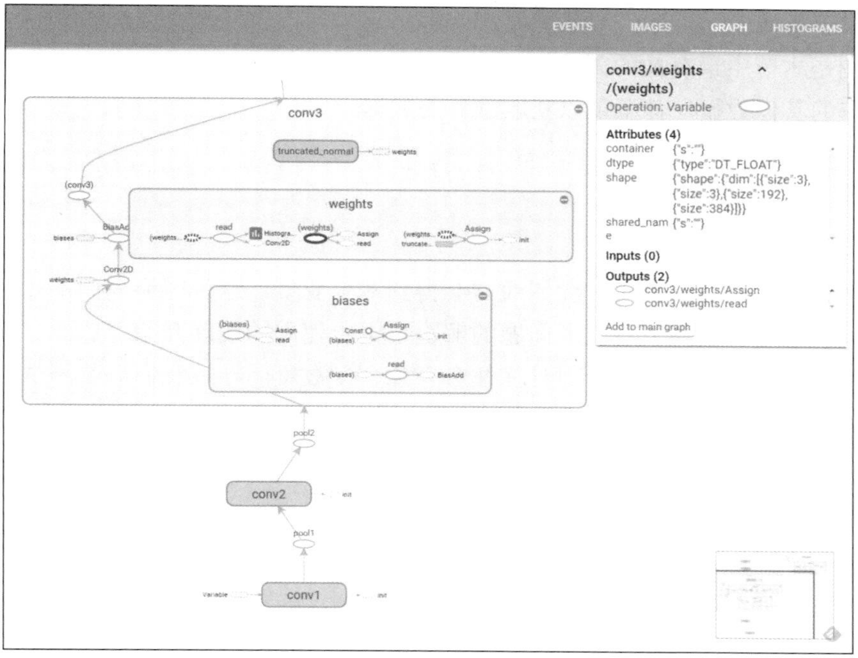

图 3-10

我们已经深入 5 层网络中的第 3 层卷积层，下面就来看一下该层的权重和偏差。

单击计算图中的权重节点，我们可以看到此类结构的细节信息：`{"shape":{"dim":[{"size":3},{"size":3},{"size":192},{"size":384}]}}`。我们可以将这些细节信息与原始论文和之前引用的图形相对比，还可以在代码中追溯网络设置的细节。

```
with tf.name_scope('conv3') as scope:
  kernel = tf.Variable(tf.truncated_normal([3, 3, 192, 384],
                       dtype=tf.float32,
                       stddev=1e-1), name='weights')
  conv = tf.nn.conv2d(pool2, kernel, [1, 1, 1, 1],
   padding='SAME')
  biases = tf.Variable(tf.constant(0.0, shape=[384],
   dtype=tf.float32),
                       trainable=True, name='biases')
  bias = tf.nn.bias_add(conv, biases)
  conv3 = tf.nn.relu(bias, name=scope)
  parameters += [kernel, biases]
```

图形浏览器和代码中的细节是一样的，但是使用 TensorBoard 很容易将数据流可视化，也很容易折叠重复部分，以及展开感兴趣的部分，如图 3-10 所示。

计算图是本节中最有趣的部分。我们也可以运行修改后的脚本并查看训练效果，还可以查看获取的其他数据，甚至可以获取额外的数据。

3.3 自动化运行

当试图训练一个分类器时，我们通常会得到多个变量，这是因为我们不知道如何设置才是好的。本节，我们从查看类似问题的解决方案所使用的设置开始。然而，我们通常会得到一系列有待测试的可能值，而且往往会有几个参数可能导致我们需要测试更多组合，从而使测试变得更复杂。

对于这种情况，建议将我们感兴趣的参数保留为可能传递到训练器中的值。wrapper 脚本可以输入参数的不同组合，并包含一个唯一的输出日志子目录，该子目录可能被用一个描述性的名称标记。

这使得我们可以轻易地比较多个测试的结果值和中间值。图 3-11 同时给出了 3 个运行状态的损失值。从中可以轻易地看到表现不佳和表现出众的运行状态。

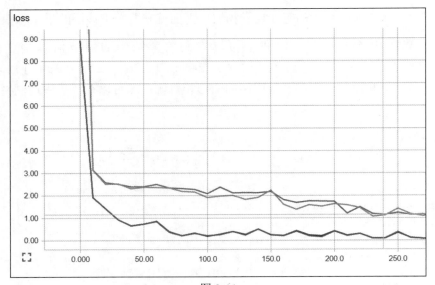

图 3-11

3.4 总结

在本章中,我们学习了 TensorBoard 的主要区域——**EVENTS**、**HISTOGRAMS** 和 **GRAPH**。我们修改了流行的模型,以了解 TensorBoard 启动和运行所需要的准确修改,这证明了使用 TensorBoard 仅仅需要做相当少量的工作。

我们还查看了各种流行模型的网络设计。我们将 TensorBoard 嵌入到代码中,并使用 TensorBoard 图形浏览器深入分析了网络设置。

读者现在应该能够更高效地使用 TensorBoard,评估训练性能,计划运行状态和修改训练脚本。

接下来,我们将学习卷积网络。我们将使用之前工作的部分成果,这样就可以马上开始工作。后文我们将关注更高级的神经网络设置,以获取更高的准确率。对训练准确率的关注反映了大多数从业者努力的焦点,是时候面对挑战了。

第 4 章 猫和狗

在第 2 章中，我们构建了一个简单的神经网络用于字符识别，并最终取得了 85% 左右的准确率。

在本章中，我们将利用更强大的网络架构来改进之前的分类器；然后，将深入研究一个更加困难的问题：处理 CIFAR-10 数据集中的彩色图像。这些图像非常复杂（包含猫、狗、飞机等），需要我们引入更强大的工具：卷积神经网络。

4.1 回顾 notMNIST

针对第 2 章使用的 notMNIST 数据集，本章我们将开始逐步改善所使用的技术。读者可以在学习本章过程中逐步编写代码，或者直接使用本书代码库中的代码（第 2 章的 training 文件）。

首先，导入下面的工具包。

```
import sys, os
import tensorflow as tf
sys.path.append(os.path.realpath('../..'))
from data_utils import *
from logmanager import *
import math
```

此处虽然并没有太大的变化，但是真正的"利器"已经通过 tensorflow 包导入。你可能会注意到，我们再次使用了之前创建的 data_utils，但是需要做一些修改。

与之前唯一的区别是 math 包，我们将使用其中的一些数学函数，例如 ceiling。

4.1.1 程序配置

之前的程序配置如下所示。

```
batch_size = 128
num_steps = 10000
learning_rate = 0.3
data_showing_step = 500
```

但是这次我们需要更多的配置。下面是将要使用的配置。

```
batch_size = 32
num_steps = 30000
learning_rate = 0.1
data_showing_step = 500
model_saving_step = 2000
log_location = '/tmp/alex_nn_log'

SEED = 11215

patch_size = 5
depth_inc = 4
num_hidden_inc = 32
dropout_prob = 0.8
conv_layers = 3
stddev = 0.1
```

前面 4 个配置是我们比较熟悉的。

- 我们仍将进行一定数量步骤（num_steps）的训练，不过此处步骤的数量增大了。随着数据集变得更加复杂、需要更多的训练，步骤数量甚至会变得更大。

- 后续我们将回顾学习率（learning_rate）的微妙之处，但首先需要确保你已经对它很熟悉了。

- 我们将每隔 500 步就检查一次中间结果，这由 data_showing_step 变量控制。

- 最后，log_location 变量控制 TensorBoard 日志转储的位置。可以参考第 3 章的相应内容。

接下来配置**随机种子**（SEED）变量，也可以不设置它，那么 TensorFlow 将会在每次运行时取一个随机数。将不同运行状态中的种子变量设置为同一个常数，那么我们调试系统时就能够保证多次运行的一致性。如果要使用种子变量，并且确实需要通过它来开始，那么可以将它设置成任何你喜欢的数字，例如你的生日、周年纪念日、第一个电话号码或者幸运数字，而我使用了我钟爱的街区的邮政编码。

最后，我们将会遇到 7 个新变量：`batch_size`、`patch_size`、`depth_inc`、`num_hidden_inc`、`conv_layers`、`stddev` 和 `dropout_prob`。它们是**卷积神经网络**（**Convolutional Neural Networks**，**CNNs**）工作原理的核心，当我们探索正在使用的网络时，将会在上下文引入它们。

4.1.2 理解卷积神经网络

在机器学习领域，CNNs 是专门用于处理图像的更高级的神经网络。不同于我们之前使用的隐藏层，CNNs 拥有一些不完全连接的层。这些层除了具有宽度和高度外，还具有深度。图像处理的普遍原则是逐块对图像进行分析。在图 4-1 中，我们可以看到一个 7 像素×7 像素块。

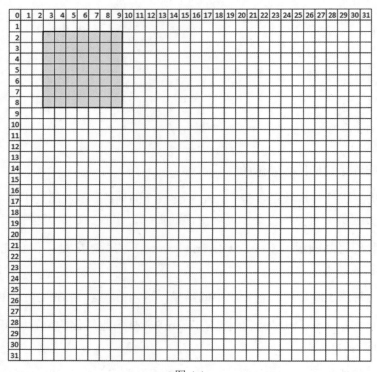

图 4-1

图 4-1 展示了一个 32 像素×32 像素的灰度图像，使用了一个 7 像素×7 像素的图像块。从左向右滑动图像块的示例如图 4-2 所示。

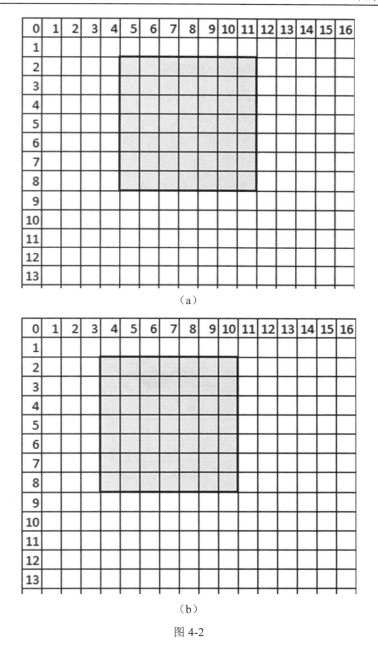

图 4-2

如果这是一个彩色图像，那么我们会同时在 3 个相同的层上滑动图像块。

你可能注意到，我们是逐像素地滑动图像块。我们也可以每次滑动过更多像素，例如每次 2 像素甚至 3 像素，这就是步幅配置。步幅越大，我们覆盖的图像就会越少，输出层就越小。

矩阵数学可以将图像块由通道数量驱动的全深度减小到输出深度列的大小,但此处不作详述。输出在高度和宽度上只有一个像素,但在深度上具有很多像素。由于我们会在迭代过程中反复滑动图像块,因此深度列的序列将会构成一个具有新长度、新宽度和新高度的块。

这里还有另一个配置——沿着图像边界填充。填充越多,那么图像块滑动和转向图像边缘的空间就越多。这就允许更多种类的步幅,从而使输出结果具有更多种类的长度和宽度。我们将会在后续的代码中看到这一点(代码中设置 `padding='SAME'` 或者 `padding='VALID'`)。

接下来看看这些图像块是如何叠加的。首先选择一个图像块,如图 4-3 所示。

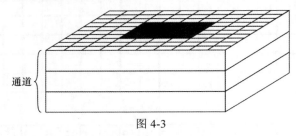

图 4-3

该图像块不仅是正方体,而且是全深度的(对于彩色图像而言),如图 4-4 所示。

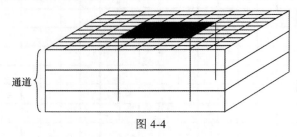

图 4-4

然后将它转换为一个具有深度的 1×1 卷,如图 4-5 所示。转换生成的卷的深度是可配置的,我们在程序中将使用 `inct_depth` 来配置该深度。

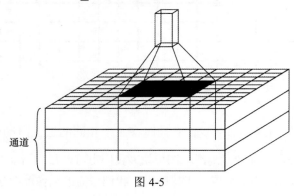

图 4-5

最后，当我们将该图像块反复滑过原始图像时，将会产生很多这种 $1 \times 1 \times N$ 的卷，图像块本身也产生了一个卷，如图 4-6 所示。

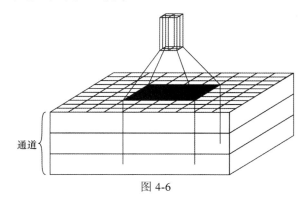

图 4-6

然后我们将它转换成一个 1×1 的卷。

最后，我们将使用池化（POOL）操作来压缩结果卷的每一层，如图 4-7 所示。池化操作有很多种类型，但比较典型的是**最大池化**（**max pooling**）。

	0	1	2	3	4	5	6	7
0	0.5	0.63	0.35	0.46	0.63	0.24	0.7	0.95
1	0.56	0.3	0.32	0.52	0.29	0.95	0.76	0.1
2	0.33	0.15	0.42	0.01	0.61	0.33	0.66	0.74
3	0.9	0.73	0.22	0.16	0.81	0.74	0.21	0.67
4	0.92	0.83	0.02	0.67	0.97	0.32	0.6	0.11
5	0.86	0.81	0.3	0.83	0.78	0.97	0.86	0.35
6	0.8	0.92	0.65	0.39	0.16	0.45	0.66	0.89
7	0.39	0.96	0.12	0.02	0.26	0.73	0.4	0.53

图 4-7

与之前使用的滑动图像块相同，池化操作也会存在一个块（这次除外，我们将使用最大数量的块）和一个步幅（这次我们想使用一个更大的步幅来压缩图像）。本质上我们是在减小尺寸。这里，我们将使用一个 3×3 的块，其步幅为 2。

4.1.3　回顾配置

我们已经介绍了卷积神经网络，现在来回顾一下之前遇到的配置：`batch_size`、`patch_size`、`depth_inc`、`num_hidden_inc`、`conv_layers`、`stddev` 和 `dropout_prob`。

- 批量尺寸（`batch_size`）。
- 块大小（`patch_size`）。
- 深度增量（`depth_inc`）。
- 隐藏增量的数量（`num_hidden_inc`）。
- 卷积层（`conv_layers`）。
- 标准差（`stddev`）。
- 失活（dropout）概率（`dropout_prob`）。

4.1.4　构造卷积神经网络

此处将跳过对 `reformat` 和 `accuracy` 两个功能函数的解释，因为我们在第 2 章中已经介绍过它们。我们将直接跳到神经网络的配置部分。为了进行比较，图 4-8 显示了第 2 章中的模型，图 4-9 显示了我们的新模型。接下来将会在相同的 notMNIST 数据集上运行新模型，以查看将会取得的准确性的提升效果。

```
def nn_model(data, weights, biases):
    layer_fc1 = tf.matmul(data, weights['fc1']) + biases['fc1']
    relu_layer = tf.nn.relu(layer_fc1)
    for relu in range(2, relu_layers + 1):
        relu_layer = tf.nn.relu(relu_layer)
    return tf.matmul(relu_layer, weights['fc2']) + biases['fc2']
```

图 4-8

首先，我们将使用一个辅助函数，如下所示。

```
def fc_first_layer_dimen(image_size, layers):
  output = image_size
  for x in range(layers):
   output = math.ceil(output/2.0)
  return int(output)
```

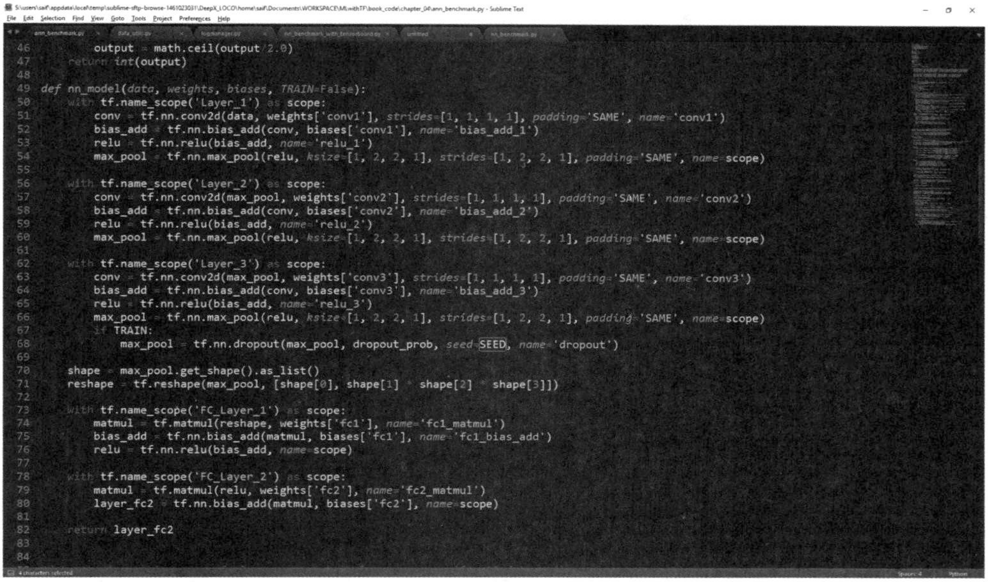

图 4-9

然后是我们稍后要调用的函数,如下所示。

```
fc_first_layer_dimen(image_size, conv_layers)
```

`fc_first_layer_dimen` 函数计算第一个全连接层的维度。CNNs 的典型做法是利用一个更小的窗口来逐层使用一系列网络层。在这里,我们决定将使用的每个卷积层的维度减半。这也说明了当输入图像的维度能被 2 的幂次方整除时事情会变得简单的原因。

现在来解析实际的网络。该网络是使用 `nn_model` 方法生成的,并且在后续训练模型时将会被调用,在对验证集和测试集进行测试时也会被再次调用。

回想一下 CNNs 通常是如何由以下几层组成的。

◆ 卷积层。

◆ 修正线性单元层(ReLU 层)。

◆ 池化层。

◆ 全连接层。

卷积层通常与 RELU 层成对重复出现。我们曾得到 3 对几乎完全相同的卷积-ReLU(CONV-RELU)层,相互堆叠在彼此的顶部。

每一个成对的层看起来都如下所示。

```
with tf.name_scope('Layer_1') as scope:
    conv = tf.nn.conv2d(data, weights['conv1'], strides=[1, 1,
        1, 1], padding='SAME', name='conv1')
    bias_add = tf.nn.bias_add(conv, biases['conv1'],
        name='bias_add_1')
    relu = tf.nn.relu(bias_add, name='relu_1')
    max_pool = tf.nn.max_pool(relu, ksize=[1, 2, 2, 1],
        strides=[1, 2, 2, 1], padding='SAME', name=scope)
```

这 3 个几乎完全相同的层（Layer_1、Layer_2 和 Layer_3）之间的主要区别是，某层的输出如何连续地输入下一层。第一层的输入是数据（图像数据），而第二层的输入是来自第一层的池化层的输出，如下所示。

```
conv = tf.nn.conv2d(max_pool, weights['conv2'], strides=[1, 1, 1,
    1], padding='SAME', name='conv2')
```

类似地，第三层的输入是第二层的池化层的输出，如下所示。

```
conv = tf.nn.conv2d(max_pool, weights['conv3'], strides=[1, 1, 1,
    1], padding='SAME', name='conv3')
```

在这 3 个 CONV-RELU 层之间，还存在另一个主要区别，那就是层压缩的情况不同。使用 print 语句来声明每一层有助于查看 conv 变量，如下所示。

```
print "Layer 1 CONV", conv.get_shape()
print "Layer 2 CONV", conv.get_shape()
print "Layer 3 CONV", conv.get_shape()
```

结构显示如下。

```
Layer 1 CONV (32, 28, 28, 4)
Layer 2 CONV (32, 14, 14, 4)
Layer 3 CONV (32, 7, 7, 4)
Layer 1 CONV (10000, 28, 28, 4)
Layer 2 CONV (10000, 14, 14, 4)
Layer 3 CONV (10000, 7, 7, 4)
Layer 1 CONV (10000, 28, 28, 4)
Layer 2 CONV (10000, 14, 14, 4)
Layer 3 CONV (10000, 7, 7, 4)
```

我们利用 notMNIST 数据集来运行它，毫无意外，我们会看到原始层的输入大小为 28×28，而连续卷积层的大小分别为 14×14 和 7×7。注意观察连续卷积层的过滤器是如何被压缩的。

为了让事情变得更有趣，我们来检查整个网络栈。添加以下 print 语句来观察 CONV 层、RELU 层和 POOL 层。

```
print "Layer 1 CONV", conv.get_shape()
print "Layer 1 RELU", relu.get_shape()
print "Layer 1 POOL", max_pool.get_shape()
```

在另外两个 CONV-RELU-POOL 网络栈后面添加类似的语句，将会得到以下输出。

```
Layer 1 CONV (32, 28, 28, 4)
Layer 1 RELU (32, 28, 28, 4)
Layer 1 POOL (32, 14, 14, 4)
Layer 2 CONV (32, 14, 14, 4)
Layer 2 RELU (32, 14, 14, 4)
Layer 2 POOL (32, 7, 7, 4)
Layer 3 CONV (32, 7, 7, 4)
Layer 3 RELU (32, 7, 7, 4)
Layer 3 POOL (32, 4, 4, 4)
...
```

我们将忽略验证和测试实例（它们是相同的，除了我们在处理验证集和测试集时的高度是 10 000，而不是最小批次 32）的输出。

从输出中我们可以看到维度（28～14）是如何在池化层中被压缩的，以及压缩后它是如何传递到下一个卷积层的。在第三个也就是最后一个池化层中，会得到一个 4×4 大小的输出。

最后的卷积层还存在另一个特点——一个失活层，训练数据时将会用到这一层，如下所示。

```
max_pool = tf.nn.dropout(max_pool, dropout_prob, seed=SEED,
 name='dropout')
```

该层利用了之前的配置 dropout_prob = 0.8。它利用 dropouts 禁止该层上相邻节点间协同工作来随机删除神经元，以此防止过度拟合；它们永远不能依赖于一个现存的特定节点。

下面继续操作网络。我们会发现一个全连接层，紧接着是一个 ReLU 层。

```
with tf.name_scope('FC_Layer_1') as scope:
    matmul = tf.matmul(reshape, weights['fc1'],
     name='fc1_matmul')
    bias_add = tf.nn.bias_add(matmul, biases['fc1'],
     name='fc1_bias_add')
```

```
    relu = tf.nn.relu(bias_add, name=scope)
```

最后，将得到一个全连接层，如下所示。

```
with tf.name_scope('FC_Layer_2') as scope:
    matmul = tf.matmul(relu, weights['fc2'],
      name='fc2_matmul')
    layer_fc2 = tf.nn.bias_add(matmul, biases['fc2'],
      name=scope)
```

这是卷积网络的典型特征，即通常会先得到一个全连接的 ReLU 层，最后得到一个持有所有类别分数的全连接层。

我们跳过了过程中的一些细节。大多数网络层是利用另外 3 个值（`weights`、`biases` 和 `strides`）来初始化的，如图 4-10 所示。

```
weights = {
    'conv1': tf.Variable(tf.truncated_normal(shape=[patch_size, patch_size, num_channels, depth_inc], dtype=tf.float32,
                         stddev=stddev, seed=SEED), name='weights_conv1'),
    'conv2': tf.Variable(tf.truncated_normal([patch_size, patch_size, depth_inc, depth_inc], dtype=tf.float32,
                         stddev=stddev, seed=SEED), name='weights_conv2'),
    'conv3': tf.Variable(tf.truncated_normal([patch_size, patch_size, depth_inc, depth_inc], dtype=tf.float32,
                         stddev=stddev, seed=SEED), name='weights_conv3'),
    'fc1': tf.Variable(tf.truncated_normal([(fc_first_layer_dimen(image_size, conv_layers) ** 2) * depth_inc,
                         num_hidden_inc], dtype=tf.float32,
                         stddev=stddev, seed=SEED), name='weights_fc1'),
    'fc2': tf.Variable(tf.truncated_normal([num_hidden_inc, num_of_classes], dtype=tf.float32,
                         stddev=stddev, seed=SEED), name='weights_fc2')
}
biases = {
    'conv1': tf.Variable(tf.zeros(shape=[depth_inc], dtype=tf.float32), name='biases_conv1'),
    'conv2': tf.Variable(tf.zeros(shape=[depth_inc], dtype=tf.float32), name='biases_conv2'),
    'conv3': tf.Variable(tf.zeros(shape=[depth_inc], dtype=tf.float32), name='biases_conv3'),
    'fc1': tf.Variable(tf.zeros(shape=[num_hidden_inc], dtype=tf.float32), name='biases_fc1'),
    'fc2': tf.Variable(tf.zeros(shape=[num_of_classes], dtype=tf.float32), name='biases_fc2')
}
```

图 4-10

`weights` 和 `biases` 本身是用其他变量初始化的，但并不是说这样就会很容易。

这里最重要的变量是 `patch_size`，它表示滑过图像时的过滤器大小。我们之前将它设置为 5，所以将使用 5×5 的块。后续我们还会重新介绍 `stddev` 和 `depth_inc` 的配置。

4.1.5 实现

现在，你的脑海中肯定萦绕着很多问题——为什么是 3 个卷积层，而不是 2 个或 4 个？为什么步幅为 1 呢？为什么一个块的大小是 5？为什么最后是一个全连接层，而开始时不是全连接层呢？

这里有一个疯狂的方法。CNNs 的核心是围绕图像处理构建的，而块是围绕着要寻找的特征构建的。为什么有些配置可以很好地工作，而另一些则不能，尽管一般规则确实跟直觉一致，但这一点目前还没有很好的解释。精准的网络架构是通过成千上万次的试验和错误才逐渐被发现、打磨，并逐步接近完美的，这仍旧是一项研究级别的任务。

实践者的一般方法是找到一个现有的、工作良好的体系结构（例如 AlexNet、GoogLeNet、ResNet），并对它们进行调整，以用于特定的数据集。这就是我们前面所做的，即开始时使用 AlexNet，然后对它进行调整。尽管这并没有实现，但它确实有效，并且这一做法在 2016 年一度盛行。

4.2　训练日

看到训练付诸行动，以及对先前所做的工作做出的改进，我们会越来越有成就感。我们准备如下所示的训练数据集和标签。

```
tf_train_dataset = tf.placeholder(tf.float32,
shape=(batch_size, image_size, image_size,
num_channels),
name='TRAIN_DATASET')
tf_train_labels = tf.placeholder(tf.float32,
shape=(batch_size, num_of_classes),
name='TRAIN_LABEL')
tf_valid_dataset = tf.constant(dataset.valid_dataset,
name='VALID_DATASET')
tf_test_dataset = tf.constant(dataset.test_dataset,
name='TEST_DATASET')
```

然后，运行训练器，如下所示。

```
# Training computation.
logits = nn_model(tf_train_dataset, weights, biases,
True)
loss = tf.reduce_mean(
    tf.nn.softmax_cross_entropy_with_logits(logits,
     tf_train_labels))
# L2 regularization for the fully connected
parameters.
regularizers = (tf.nn.l2_loss(weights['fc1']) +
 tf.nn.l2_loss(biases['fc1']) +
 tf.nn.l2_loss(weights['fc2']) +
 tf.nn.l2_loss(biases['fc2']))
 # Add the regularization term to the loss.
 loss += 5e-4 * regularizers
 tf.summary.scalar("loss", loss)
```

这和我们在第 2 章中所做的类似。我们实例化了网络，输入了权重和偏差的初始值，并使用训练标签定义了一个损失函数。接着定义优化器,以最小化该损失函数,如下所示。

```
optimizer = tf.train.GradientDescentOptimizer
 (learning_rate).minimize(loss)
```

使用权重和偏差为验证集预测标签,并最终为训练集预测标签。

```
train_prediction = tf.nn.softmax(nn_model(tf_train_dataset,
weights, biases, TRAIN=False))
valid_prediction = tf.nn.softmax(nn_model(tf_valid_dataset,
 weights, biases))    test_prediction =
 tf.nn.softmax(nn_model(tf_test_dataset,
 weights, biases))
```

训练会话的完整代码如图 4-11 所示。

```
with tf.Session(graph=graph) as session:
    writer = tf.summary.FileWriter(log_location, session.graph)
    merged = tf.summary.merge_all()
    tf.global_variables_initializer().run()
    print("Initialized")
    for step in range(num_steps + 1):
        sys.stdout.write('Training on batch %d of %d\r' % (step + 1, num_steps))
        sys.stdout.flush()
        # Pick an offset within the training data, which has been randomized.
        # Note: we could use better randomization across epochs.
        offset = (step * batch_size) % (dataset.train_labels.shape[0] - batch_size)
        # Generate a minibatch.
        batch_data = dataset.train_dataset[offset:(offset + batch_size), :]
        batch_labels = dataset.train_labels[offset:(offset + batch_size), :]
        # Prepare a dictionary telling the session where to feed the minibatch.
        # The key of the dictionary is the placeholder node of the graph to be fed,
        # and the value is the numpy array to feed to it.
        feed_dict = {tf_train_dataset: batch_data, tf_train_labels: batch_labels}
        summary_result, _, l, predictions = session.run([merged, optimizer, loss, train_prediction], feed_dict=feed_dict)
        writer.add_summary(summary_result, step)

        if step % data_showing_step == 0:
            acc_minibatch = accuracy(predictions, batch_labels)
            acc_val = accuracy(valid_prediction.eval(), dataset.valid_labels)
            logmanager.logger.info('# %03d  Acc Train: %03.2f%% Acc Val: %03.2f%% Loss %f' % (
                step, acc_minibatch, acc_val, l))

    logmanager.logger.info("Test accuracy: %.1f%%" % accuracy(test_prediction.eval(), dataset.test_labels))
```

图 4-11

最后,运行此会话。我们将使用之前设置的 num_steps 变量,并逐块(batch_size)在训练数据集上运行。我们将加载一部分训练数据和关联的标签,然后运行该会话,如下所示。

```
batch_data = dataset.train_dataset[offset:(offset +
 batch_size), :]
batch_labels = dataset.train_labels[offset:
 (offset +
 batch_size), :]
```

得到了对小批量数据的预测结果后,使用之前声明的 valid_prediction,将它与实际的标签进行对比,以得到在小批量数据上的准确率,如下所示。

```
valid_prediction =
```

```
tf.nn.softmax(nn_model(tf_valid_dataset,
 weights, biases))
```

再根据已知的实际标签来评估对验证集的预测准确率，如下所示。

```
accuracy(valid_prediction.eval(),
dataset.valid_labels)
```

完成所有步骤之后，在测试集中做同样的操作。

```
accuracy(test_prediction.eval(), dataset.test_labels)
```

由此可知，训练集、验证集和测试集的实际执行情况与之前并没有太大的不同，不同的是准确率。需要注意的是，我们在测试集上的预测准确率已经突破 80%，并提升到 90% 以上，如图 4-12 所示。

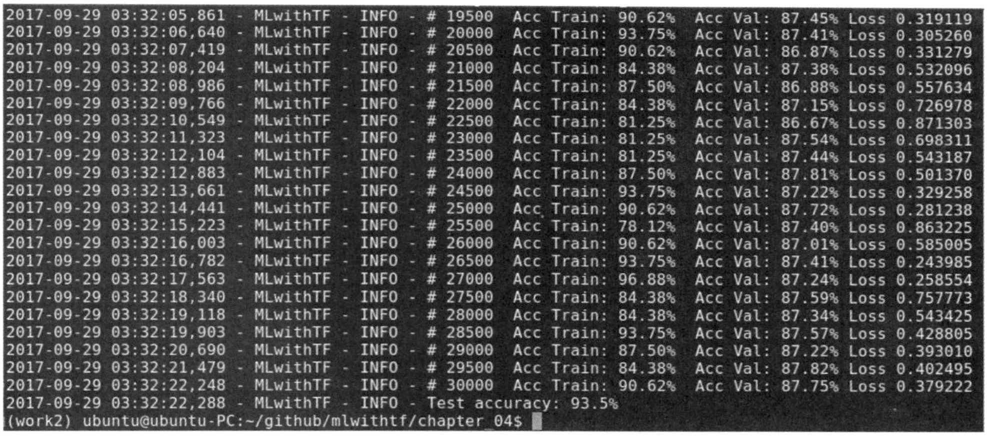

图 4-12

4.3 真实的猫和狗

我们已经在 notMNIST 数据集上演示了新工具，这个新工具对我们来说很有帮助，因为可以使用它与之前设置的简单网络进行比较。现在，我们继续解决一个更棘手的问题——真实的猫和狗。

此处将使用 CIFAR-10 数据集。该数据集不止包含猫和狗，它一共包含了 10 个类别：飞机、汽车、鸟、猫、鹿、狗、青蛙、马、轮船和卡车。与 notMNIST 数据集不同，CIFAR-10 数据集存在两个复杂的情况，如下所示。

◆ 照片中存在更多的不均匀性，包括背景、场景。

◆ 照片是彩色的。

我们之前没有使用过彩色数据集。幸运的是，它和通常的黑白数据集并没有太大的不同，只需添加另一个维度。例如，之前的 28×28 图像是平面矩阵，而现在我们将拥有 $32 \times 32 \times 3$ 矩阵——第三个维度代表红色、绿色和蓝色通道的层。这确实使得可视化数据集变得更加困难，因为叠加图像会进入第四维空间。因此，训练集、验证集、测试集的维度将是 $32 \times 32 \times 3 \times SET\_SIZE$。我们需要习惯面对那些无法在熟悉的 3D 空间中可视化的矩阵。

颜色维度的机制是相同的。与之前使用浮点数表示不同色度的灰色相同，我们也使用浮点数表示红色、绿色和蓝色的色度。

加载 notMNIST 数据集的方式如下所示。

```
dataset, image_size, num_of_classes, num_channels =
 prepare_not_mnist_dataset()
```

其中，`num_channels` 变量表示颜色通道数，且通道数仅为 1。

我们以类似的方式加载 CIFAR-10 数据集，这次将返回 3 个通道，如下所示。

```
dataset, image_size, num_of_classes, num_channels =
 prepare_cifar_10_dataset()
```

注意不要重新造轮子。

根据第 2 章中自动获取、提取和准备 notMNIST 数据集的方法，即将管道函数放到 `data_utils.py` 文件中，从而将管道代码与实际的机器学习代码分开。清晰的分离和整洁、通用的功能，使得我们能够在当前的项目中再次使用那些代码。

我们将再次使用第 2 章的 9 个函数，如下所示。

◆ `download_hook_function`。

◆ `download_file`。

◆ `extract_file`。

◆ `load_class`。

◆ `make_pickles`。

◆ `randomize`。

- make_arrays。
- merge_datasets。
- pickle_whole。

此时需要考虑如何在全局函数 prepare_not_mnist_dataset 中使用这些函数。该全局函数运行了整个管道，这为我们节省了不少时间。

为 CIFAR-10 数据集创建一个类似的函数。通常来说，你应该保存你自己的管道函数，并尝试将它们泛化，然后将它们隔离到单个模块中，并在不同项目中再次使用它们。这样当你在做自己的项目时，你可以将重点放在关键的机器学习工作上，而不是将时间花费在重建管道上。

注意 data_utils.py 修改后的版本，全局函数 prepare_cifar_10_dataset 将数据集细节和为该数据集准备的管道隔离开来，如下所示。

```
def prepare_cifar_10_dataset():
  print('Started preparing CIFAR-10 dataset')
  image_size = 32
  image_depth = 255
  cifar_dataset_url = 'https://www.c****/~kriz/cifar-
    10-python.tar.gz'
  dataset_size = 170498071
  train_size = 45000
  valid_size = 5000
  test_size = 10000
  num_of_classes = 10
  num_of_channels = 3
  pickle_batch_size = 10000
```

下面是对上述代码的快速概览。

- 使用 cifar_dataset_url ='https://www.c****/~kriz/cifar-10-python.tar.gz'从多伦多大学 Alex Krizhevsky 的网站上获取数据集（该数据集可从异步社区获取）。

- 使用 dataset_size=170498071 来验证是否成功接收到了文件，避免文件下载一半被截断。

- 根据对数据集的了解来声明一些细节。

- 将数据集中的 60 000 张图片分割成训练集、验证集和测试集，其大小分别对应为

45 000、5 000 和 10 000 张图片。

- 因为数据集中包含 10 个类别的图片，所以设置 `num_of_classes = 10`。
- 因为这些图片为包含红色、绿色和蓝色 3 个通道的彩色图片，所以设置 `num_of_channels = 3`。
- 因为图像是 32 像素×32 像素的，所以设置 `image_size=32`，并将该值同时用于宽度和高度。
- 图像在每个通道上都是 8 位，因此设置 `image_depth = 255`。
- 数据将被保存到 /datasets/CIFAR-10/ 路径下。

与我们对 notMNIST 数据集所做的非常类似，只有在没有该数据集的情况下才会下载它。我们将解压数据集、执行必要的转换，并使用 `pickle_cifar_10` 函数将预处理后的矩阵保存为 pickle 文件。如果找到了 pickle 文件，可以使用 `load_cifar_10_from_pickles` 方法重新加载中间数据。

我们使用下面的 3 个辅助方法来确保主方法的复杂性是可控的。

- `pickle_cifar_10`。
- `load_cifar_10_from_pickles`。
- `load_cifar_10_pickle`。

函数的定义如图 4-13 所示。

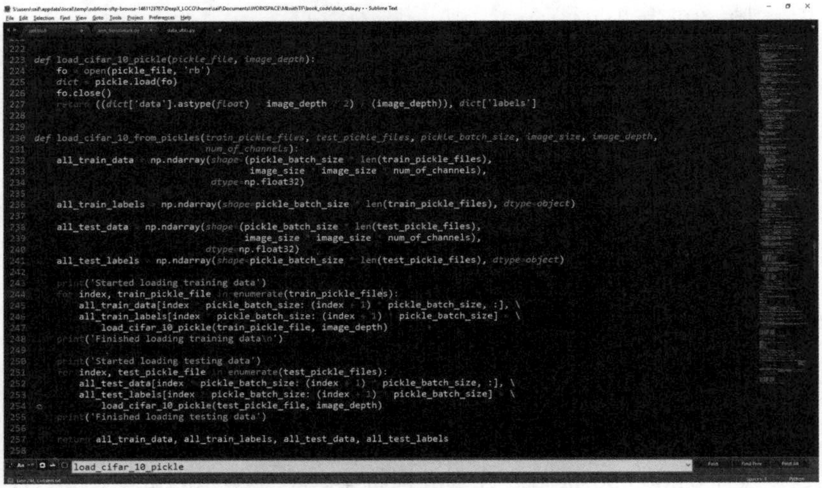

图 4-13

load_cifar_10_pickle 方法用于为训练数据、测试数据和标签分配 numpy 数组，并将存在的 pickle 文件加载到这些数组中。（因为每件事情都需要做两次，所以将 load_cifar_10_pickle 方法隔离出来，实际上该方法会加载数据并对其进行零中心化（zero-center）处理。）代码如图 4-14 所示。

```
def pickle_cifar_10(all_train_data, all_train_labels, all_test_data, all_test_labels,
                    train_size, valid_size, test_size, output_file_path, FORCE=False):

    if os.path.isfile(output_file_path) and not FORCE:
        print('Pickle file: %s already exist' % output_file_path)

        with open(output_file_path, 'rb') as f:
            save = pickle.load(f)
            train_dataset = save['train_dataset']
            train_labels = save['train_labels']
            valid_dataset = save['valid_dataset']
            valid_labels = save['valid_labels']
            test_dataset = save['test_dataset']
            test_labels = save['test_labels']
            del save  # hint to help gc free up memory
            print('Training set', train_dataset.shape, train_labels.shape)
            print('Validation set', valid_dataset.shape, valid_labels.shape)
            print('Test set', test_dataset.shape, test_labels.shape)

        return train_dataset, train_labels, valid_dataset, valid_labels, test_dataset, test_labels
    else:
        train_dataset = all_train_data[0:train_size]
        train_labels = all_train_labels[0:train_size]
        valid_dataset = all_train_data[train_size:train_size + valid_size]
        valid_labels = all_train_labels[train_size:train_size + valid_size]
        test_dataset = all_test_data[0:test_size]
        test_labels = all_test_labels[0:test_size]

        try:
            f = open(output_file_path, 'wb')
            save = {
                'train_dataset': train_dataset,
                'train_labels': train_labels,
                'valid_dataset': valid_dataset,
                'valid_labels': valid_labels,
                'test_dataset': test_dataset,
                'test_labels': test_labels,
            }
            pickle.dump(save, f, pickle.HIGHEST_PROTOCOL)
            f.close()
        except Exception as e:
            print('Unable to save data to', output_file_path, ':', e)
            raise

        statinfo = os.stat(output_file_path)
        print('Compressed pickle size:', statinfo.st_size)

    return train_dataset, train_labels, valid_dataset, valid_labels, test_dataset, test_labels
```

图 4-14

同样，我们将检查是否已经存在对应的 pickle 文件，如果存在，就加载它们。只有当它们不存在（对应 else 子句）时，才会将准备的数据保存到 pickle 文件。

4.4 保存模型以供持续使用

为了保存 TensorFlow 会话中的变量以备将来使用，可以使用 Saver() 函数。如下所示，我们在 writer 变量之后紧接着创建一个 saver 变量。

```
writer = tf.summary.FileWriter(log_location, session.graph)
saver = tf.train.Saver(max_to_keep=5)
```

然后，在训练循环中添加以下代码，实现每隔 `model_saving_step` 步就保存一次模型。

```
if step % model_saving_step == 0 or step == num_steps + 1:
  path = saver.save(session, os.path.join(log_location,
"model.ckpt"), global_step=step)
  logmanager.logger.info('Model saved in file: %s' % path)
```

之后，无论何时想恢复保存的模型，都可以通过创建一个新的 `Saver()` 实例，并使用 `restore` 函数来轻松地实现，如下所示。

```
checkpoint_path = tf.train.latest_checkpoint(log_location)
restorer = tf.train.Saver()
with tf.Session() as sess:
    sess.run(tf.global_variables_initializer())
    restorer.restore(sess, checkpoint_path)
```

在上面的代码中，我们使用了 `tf.train.latest_checkpoint`，这样 TensorFlow 将会自动选择最新的模型检查点。然后，我们创建了一个名为 `restore` 的 `Saver()` 新实例。最后，使用 `restore` 函数将所保存的模型加载到会话图中。

```
restorer.restore(sess, checkpoint_path)
```

需要注意的是，必须在运行 `tf.global_variables_initializer` 之后才能恢复模型，否则，已经加载的变量将会被初始化程序覆盖。

4.5 使用分类器

现在我们已经增强了分类器来加载随机图片，接下来将开始以训练集或测试集图片的准确大小和形状来选择这些随机图片。需要为这些图片添加占位符，因此将在合适的位置添加以下代码。

```
tf_random_dataset = tf.placeholder(tf.float32, shape=(1,
 image_size, image_size, num_channels),
name='RANDOM_DATA')random_prediction =
tf.nn.softmax(nn_model(tf_random_dataset,
 weights, biases))
```

接下来，通过图 4-15 所示的命令行参数获取用户提供的图像，并在图像上运行会话。

```python
if (evaluateFile is not None):
    image = (ndimage.imread(evaluateFile).astype(float) - 255 / 2) / 255
    image = image.reshape((image_size, image_size, num_channels)).astype(np.float32)
    random_data = np.ndarray((1, image_size, image_size, num_channels), dtype=np.float32)
    random_data[0, :, :, :] = image

    feed_dict = {tf_random_dataset: random_data}
    output = session.run(
        [random_prediction], feed_dict=feed_dict)

    for i, smx in enumerate(output):
        prediction = smx[0].argmax(axis=0)
        print 'The prediction is: %d' % (prediction)
```

图 4-15

我们将严格按照之前的顺序来操作。通过脚本使用 -e 开关来运行一个 test 文件，这将会产生一个额外的输出，如下所示。

```
The prediction is: 2
```

我们对任意一个图像都进行了分类。

4.6　所学技能

在本章，你应该学到了下面这些技能。

◆ 准备更高级的彩色图像训练数据和测试数据。

◆ 建立一个卷积神经网络。

◆ 与 CNNs 相关的参数和配置。

◆ 创建一个包含 TensorBoard hook 的完整系统。

◆ 真实数据处理流程。

4.7　总结

在本章中，我们创建了一个更高级的分类器，实现了导入导出模型，并开始将分类器应用到任意模型中。此外，我们还训练了一个系统来区分猫和狗。

在下一章中，我们将开始使用序列到序列模型，并利用 TensorFlow 编写一个英语到法语的翻译器。

第 5 章
序列到序列模型——你讲法语吗

到目前为止，我们的大部分工作是关于图像的。图像处理在表现机器学习的快速性和简洁性方面有很重要的作用。然而，机器学习领域的应用范围远不止于此，接下来的几章将会涵盖机器学习在其他方面的应用。我们将从序列到序列模型开始。尽管它的设置有些复杂，而且训练数据集要大得多，但是其结果同样令人惊讶。

在本章中，我们将重点关注以下几个方面。

- 理解序列到序列模型（sequence-to-sequence model）是如何工作的。
- 理解序列到序列模型所需要的设置。
- 使用序列到序列模型创建一个将英语翻译成法语的翻译器。

5.1 快速预览

我们将会编写一个将英语翻译成法语的翻译器。之前的机器学习方法可能会通过一系列的解析器和规则来解决该问题，规则限定了如何将词汇与短语翻译成其他语言，但现在我们的方法将更加优雅、通用和快速。我们将使用很多示例来训练我们的翻译器。

我们要做的是建立一个包含足够多的将英语句子翻译成法语句子（实际上，针对任何语言都能工作）的数据集。翻译过的文章和新闻报道没有什么作用，因为我们不一定能够把特定的文本逐行地从一种语言转换成另一种语言。所以，我们需要更有创造性的方法。例如，联合国组织经常需要逐行翻译以满足不同选区的需要。这为我们提供了便利。

在 2010 年召开的统计机器翻译研讨会（Workshop on Statistical Machine Translation）发

布了一个可用的封装训练集。

我们将使用法语对应的特定文件，文件可在异步社区下载。

以下是源数据的一段英文摘要。

- Food, where European inflation slipped up
- The skyward zoom in food prices is the dominant force behind the speed up in eurozone inflation
- November price hikes were higher than expected in the 13 eurozone countries,with October's 2.6 percent yr/yr inflation rate followed by 3.1 percent in November, the EU's Luxembourg-based statistical office reported
- Official forecasts predicted just three percent, Bloomberg said
- As opposed to the US, UK, and Canadian central banks, the **European Central Bank (ECB)** did not cut interest rates, arguing that a rate drop combined with rising raw material prices and declining unemployment would trigger an inflationary spiral
- The ECB wants to hold inflation to under two percent, or somewhere in that vicinity
- According to one analyst, ECB has been caught in a Catch-22, and it needs to **talk down** inflation, to keep from having to take action to push it down later in the game

下面是对应的法语摘要。

- L'inflation, en Europe, a dérapé sur l'alimentation
- L'inflation accélérée, mesurée dans la zone euro, est due principalement à l'augmentation rapide des prix de l'alimentation
- En novembre, l'augmentation des prix, dans les 13 pays de la zone euro, a été plus importante par rapport aux prévisions, après un taux d'inflation de 2,6 pour cent en octobre, une inflation annuelle de 3,1 pour cent a été enregistrée, a indiqué le bureau des statistiques de la Communauté Européenne situé à Luxembourg
- Les prévisions officielles n'ont indiqué que 3 pour cent, a communiqué Bloomberg
- Contrairement aux banques centrales américaine, britannique et canadienne, la Banque centrale européenne (BCE) n'a pas baissé le taux d'intérêt directeur en disant que la diminution des intérêts, avec la croissance des prix des matières premières et la baisse

du taux de chômage, conduirait à la génération d'une spirale inflationniste
- La BCE souhaiterait maintenir le taux d'inflation au-dessous mais proche de deux pour cent
- Selon un analyste, c'est le Catch 22 pour la BCE-: "il faut dissuader" l'inflation afin de ne plus avoir à intervenir sur ce sujet ultérieurement

在可能的情况下，快速进行完整性检查通常是一种不错的做法，以确保文件的顺序正确。两个文件的第 7 项都有"Catch 22"短语，这给了我们些许安慰。

对于一个统计方法来说，7 项内容远远不够。我们要实现的是一个优雅、通用的解决方案，这个方案需要大量的数据。因此，我们用作训练集的数据包含 20GB 大小的文本，并像前面的摘要那样进行了逐行翻译。

与处理图像的操作类似，我们将使用子集进行训练、验证和测试。我们还将定义一个损失函数，并尝试将损失函数值最小化。接下来，我们从数据开始。

5.2 大量信息

按照前面的做法，你可以从异步社区获取代码。

我们重点关注 chapter_05 子文件夹，它包含以下 3 个文件：
- `data_utils.py`；
- `translate.py`；
- `seq2seq_model.py`。

第一个文件用来处理数据，因此我们从该文件开始。`prepare_wmt_dataset` 函数可实现具体的操作内容，它和我们过去获取图像数据集的方式非常类似，只是现在获取的是两个数据子集：
- `giga-fren.release2.fr.gz`；
- `giga-fren.release2.en.gz`。

这就是我们要关注的两种语言的数据集。即将创建的翻译器的美妙之处在于这种方法是完全通用的，因此可以据此很容易地创建一个其他语言的翻译器，例如德语或西班牙语。

图 5-1 所示的为代码的特定子集。

```python
def prepare_wmt_dataset(tokenizer=None):
    vocab_size = 40000

    # URLs for WMT data
    _WMT_ENFR_TRAIN_URL = "http://www.statmt.org/wmt10/training-giga-fren.tar"
    _WMT_ENFR_DEV_URL = "http://www.statmt.org/wmt15/dev-v2.tgz"

    # Expected number of bytes for the above two file downloads
    _WMT_ENFR_TRAIN_SIZE = 2595102720
    _WMT_ENFR_DEV_SIZE = 21393583

    train_file_path = download_file(_WMT_ENFR_TRAIN_URL,
                                    os.path.realpath('../datasets/WMT'), _WMT_ENFR_TRAIN_SIZE)

    dev_file_path = download_file(_WMT_ENFR_DEV_URL,
                                  os.path.realpath('../datasets/WMT'), _WMT_ENFR_DEV_SIZE)

    train_extracted_folder = extract_file(train_file_path, os.path.realpath('../datasets/WMT/train'), IS_SUB=False)
    dev_extracted_folder = extract_file(dev_file_path, os.path.realpath('../datasets/WMT'), IS_SUB=False,
                                        FORCE=True)

    train_sub_gzip_files = ['giga-fren.release2.fixed.fr.gz', 'giga-fren.release2.fixed.en.gz']
    train_sub_gzip_files = [train_extracted_folder + '/' + x for x in train_sub_gzip_files]

    if not os.path.exists(train_extracted_folder + '/data'):
        os.makedirs(train_extracted_folder + '/data')

    train_sub_extracted_files = [None] * 2
    for index, train_sub_gzip_file in enumerate(train_sub_gzip_files):
        train_sub_extracted_files[index] = extract_file(train_sub_gzip_file,
                                                        train_extracted_folder + '/data/' +
                                                        train_sub_gzip_file.split('/')[-1].split('.gz')[0],
                                                        TYPE='gz',
                                                        IS_SUB=False)
```

图 5-1

接下来，我们将从头逐行运行这两个文件，并做两件事——创建词汇表和标记单个单词，这些是通过函数 `create_vocabulary` 和 `data_to_token_ids` 实现的，稍后我们将会介绍这两个函数。现在，我们为庞大的训练集 newstest2013.fr 和小型的开发集 dev/newstest2013.en 创建词汇表并对单个单词进行标记，代码如图 5-2 所示。

```python
    vocab_paths = [None] * 2
    token_paths = [None] * 2

    # Create vocabularies of the appropriate sizes and tokenizing the vocabulary
    for index, train_sub_extracted_file in enumerate(train_sub_extracted_files):
        type = train_sub_extracted_file.split('.')[-1]
        vocab_paths[index] = "%s%s.%s" % (train_extracted_folder + '/data/', 'vocab%d' % vocab_size, type)
        token_paths[index] = "%s%s.%s" % (train_extracted_folder + '/data/', 'token%d' % vocab_size, type)
        create_vocabulary(vocab_paths[index], train_sub_extracted_files[index], vocab_size, tokenizer)
        data_to_token_ids(train_sub_extracted_files[index], token_paths[index], vocab_paths[index], tokenizer)

    dev_sub_required_files = ['dev/newstest2013.fr', 'dev/newstest2013.en']
    dev_sub_required_files = [dev_extracted_folder + '/' + x for x in dev_sub_required_files]

    if not os.path.exists(dev_extracted_folder + '/dev/data'):
        os.makedirs(dev_extracted_folder + '/dev/data')

    dev_token_paths = [None] * 2
    for index, dev_sub_required_file in enumerate(dev_sub_required_files):
        type = dev_sub_required_file.split('.')[-1]
        dev_token_paths[index] = "%s%s.%s" % (dev_extracted_folder + '/dev/data/', 'token%d' % vocab_size, type)
        data_to_token_ids(dev_sub_required_files[index], dev_token_paths[index], vocab_paths[index], tokenizer)

    def wmt(): pass

    wmt.en_train_ids_path = token_paths[1]
    wmt.fr_train_ids_path = token_paths[0]
    wmt.en_dev_ids_path = dev_token_paths[1]
    wmt.fr_dev_ids_path = dev_token_paths[0]
    wmt.en_vocab_path = vocab_paths[1]
    wmt.fr_vocab_path = vocab_paths[0]

    return wmt
```

图 5-2

之前，我们使用 `create_vocabulary` 函数创建了一个词汇表。现在，我们先创建一个空的词汇表映射 `vocab = {}`，然后运行数据文件，并针对数据文件的每一行使用一个基本的分词器来创建一系列单词。（警告：不要将其与后面 ID 函数中更重要的 token 混淆。）

如果遇到一个已经存在于词汇表中的单词，请按以下方式增加该单词出现的次数。

```
vocab[word] += 1
```

否则，为该单词初始化其出现次数，如下所示。

```
vocab[word] += 1
```

一直这样进行下去，直到遍历完训练数据集中的所有行。接下来，使用 sorted(vocab, key=vocab.get, reverse=True) 对词汇表单词按频率大小进行排序。

这一步很重要，因为我们不会保留每一个单词，仅保留频率最大的前 k 个单词，其中 k 是定义的词汇表大小（在此我们将它定义为了 40 000，你可以选择不同的值并查看它是如何影响结果的），代码如图 5-3 所示。

```
def create_vocabulary(vocabulary_path, data_path, max_vocabulary_size,
                     tokenizer=None, normalize_digits=True):
    if not gfile.Exists(vocabulary_path):
        print("Creating vocabulary %s from data %s" % (vocabulary_path, data_path))
        vocab = {}
        with gfile.GFile(data_path, mode="rb") as f:
            counter = 0
            for line in f:
                counter += 1
                if counter % 100000 == 0:
                    print("  processing line %d" % counter)
                tokens = tokenizer(line) if tokenizer else basic_tokenizer(line)
                for w in tokens:
                    word = re.sub(_DIGIT_RE, b"0", w) if normalize_digits else w
                    if word in vocab:
                        vocab[word] += 1
                    else:
                        vocab[word] = 1
            vocab_list = _START_VOCAB + sorted(vocab, key=vocab.get, reverse=True)
            if len(vocab_list) > max_vocabulary_size:
                vocab_list = vocab_list[:max_vocabulary_size]
            with gfile.GFile(vocabulary_path, mode="wb") as vocab_file:
                for w in vocab_list:
                    vocab_file.write(w + b"\n")
```

图 5-3

虽然使用句子和词汇表比较直观，但这需要读者具备较强的抽象能力，即需要我们来将学习过的每个单词翻译成一个简单的整数。可以使用 sequence_to_token_ids 函数逐行处理，代码如图 5-4 所示。

我们使用 data_to_token_ids 函数将该方法应用到整个数据文件，这需要读取训练文件，逐行迭代，并运行 sequence_to_token_ids 函数以将每行中的独立单词翻译成整数，代码如图 5-5 所示。

```
def sentence_to_token_ids(sentence, vocabulary,
                          tokenizer=None, normalize_digits=True):
    if tokenizer:
        words = tokenizer(sentence)
    else:
        words = basic_tokenizer(sentence)
    if not normalize_digits:
        return [vocabulary.get(w, UNK_ID) for w in words]
    # Normalize digits by 0 before looking words up in the vocabulary.
    return [vocabulary.get(re.sub(_DIGIT_RE, b"0", w), UNK_ID) for w in words]
```

图 5-4

```
def data_to_token_ids(data_path, target_path, vocabulary_path,
                      tokenizer=None, normalize_digits=True):
    if not gfile.Exists(target_path):
        print("Tokenizing data in %s" % data_path)
        vocab, _ = initialize_vocabulary(vocabulary_path)
        with gfile.GFile(data_path, mode="rb") as data_file:
            with gfile.GFile(target_path, mode="w") as tokens_file:
                counter = 0
                for line in data_file:
                    counter += 1
                    if counter % 100000 == 0:
                        print("  tokenizing line %d" % counter)
                    token_ids = sentence_to_token_ids(line, vocab, tokenizer,
                                                     normalize_digits)
                    tokens_file.write(" ".join([str(tok) for tok in token_ids]) + "\n")
```

图 5-5

运行代码后得到两个数字数据集。我们利用两个包含数字到词汇映射关系的句子序列，将英语翻译成法语的问题，临时转换成了一组数字翻译成另一组数字的问题。

例如我们使用["Brooklyn", "has", "lovely", "homes"]，产生词汇{"Brooklyn": 1, "has": 3, "lovely": 8, "homes": 17}，那么将会得到结果[1, 3, 8, 17]。

以下是典型的文件下载实现代码。

```
ubuntu@ubuntu-PC:~/github/mlwithtf/chapter_05$: python translate.py
Attempting to download http://****/training-giga-fren.tar
File output path: /home/ubuntu/github/mlwithtf/datasets/WMT/training-giga-fren.tar
Expected size: 2595102720
File already downloaded completely!
Attempting to download http://www.****/wmt15/dev-v2.tgz
File output path: /home/ubuntu/github/mlwithtf/datasets/WMT/dev-v2.tgz
```

```
Expected size: 21393583
File already downloaded completely!
/home/ubuntu/github/mlwithtf/datasets/WMT/training-giga-fren.tar
already extracted to
/home/ubuntu/github/mlwithtf/datasets/WMT/train
Started extracting /home/ubuntu/github/mlwithtf/datasets/WMT/dev-
v2.tgz to /home/ubuntu/github/mlwithtf/datasets/WMT
Finished extracting /home/ubuntu/github/mlwithtf/datasets/WMT/dev-
v2.tgz to /home/ubuntu/github/mlwithtf/datasets/WMT
Started extracting
/home/ubuntu/github/mlwithtf/datasets/WMT/train/giga-
fren.release2.fixed.fr.gz to
/home/ubuntu/github/mlwithtf/datasets/WMT/train/data/giga-
fren.release2.fixed.fr
Finished extracting
/home/ubuntu/github/mlwithtf/datasets/WMT/train/giga-
fren.release2.fixed.fr.gz to
/home/ubuntu/github/mlwithtf/datasets/WMT/train/data/giga-
fren.release2.fixed.fr
Started extracting
/home/ubuntu/github/mlwithtf/datasets/WMT/train/giga-
fren.release2.fixed.en.gz to
/home/ubuntu/github/mlwithtf/datasets/WMT/train/data/giga-
fren.release2.fixed.en
Finished extracting
/home/ubuntu/github/mlwithtf/datasets/WMT/train/giga-
fren.release2.fixed.en.gz to
/home/ubuntu/github/mlwithtf/datasets/WMT/train/data/giga-
fren.release2.fixed.en
Creating vocabulary
/home/ubuntu/github/mlwithtf/datasets/WMT/train/data/vocab40000.fr
from
data /home/ubuntu/github/mlwithtf/datasets/WMT/train/data/giga-
fren.release2.fixed.fr
  processing line 100000
  processing line 200000
  processing line 300000
...
  processing line 22300000
  processing line 22400000
  processing line 22500000
 Tokenizing data in
  /home/ubuntu/github/mlwithtf/datasets/WMT/train/data/giga-
```

```
fren.release2.fr
 tokenizing line 100000
 tokenizing line 200000
 tokenizing line 300000
...
 tokenizing line 22400000
 tokenizing line 22500000
Creating vocabulary
/home/ubuntu/github/mlwithtf/datasets/WMT/train/data/vocab
40000.en from data
/home/ubuntu/github/mlwithtf/datasets/WMT/train/data/giga-
fren.release2.en
 processing line 100000
 processing line 200000
...
```

因为数据集处理的英文部分与前面是完全相同的，所以这里不再重复。我们将逐行读取这个庞大的文件以创建一个词汇表，并对这两个语言文件的单词逐行进行标记。

5.3 训练日

工作的关键部分是训练，它的代码在之前介绍的第二个文件 translate.py 中。前面介绍过的 `prepare_wmt_dataset` 函数是训练的起点，它创建了两个数据集，并将数据集标记为干净整洁的数字。

训练开始的代码如图 5-6 所示。

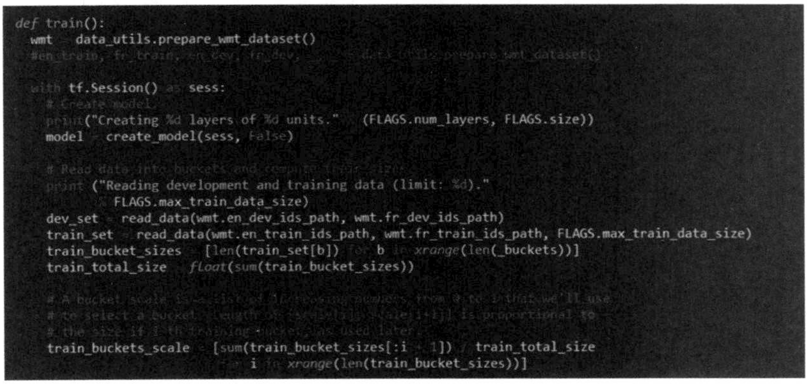

图 5-6

准备好数据之后，先创建一个 TensorFlow 会话，以此来构造我们的模型。准备工作和训练循环如下。

首先定义一个等级,它是一个取值范围为 0~1 的浮点型分数。然后训练循环,这和前几章中所做的不同,因此需要密切关注。

主要的训练循环是试图最小化误差。其中有两个关键语句,第一个关键语句如下所示。

```
encoder_inputs, decoder_inputs, target_weights =
 model.get_batch(train_set, bucket_id)
```

第二个关键语句如下所示。

```
_, step_loss, _ = model.step(sess, encoder_inputs, decoder_inputs,
 target_weights, bucket_id, False)
```

`get_batch` 函数用来将两个序列转换成主批次向量和关联的权重。将它们用在模型步骤中以返回损失值。

虽然我们不处理损失值,但会使用 `perplexity`,它是 e 的损失值次幂,如图 5-7 所示。

```
step_time, loss = 0.0, 0.0
current_step = 0
previous_losses = []
while True:
    # Choose a bucket according to data distribution. We pick a random number
    # in [0, 1] and use the corresponding interval in train_buckets_scale.
    random_number_01 = np.random.random_sample()
    bucket_id = min([i for i in xrange(len(train_buckets_scale))
                     if train_buckets_scale[i] > random_number_01])

    # Get a batch and make a step.
    start_time = time.time()
    encoder_inputs, decoder_inputs, target_weights = model.get_batch(
        train_set, bucket_id)
    _, step_loss, _ = model.step(sess, encoder_inputs, decoder_inputs,
                                 target_weights, bucket_id, False)
    step_time += (time.time() - start_time) / FLAGS.steps_per_checkpoint
    loss += step_loss / FLAGS.steps_per_checkpoint
    current_step += 1

    # Once in a while, we save checkpoint, print statistics, and run evals.
    if current_step % FLAGS.steps_per_checkpoint == 0:
        # Print statistics for the previous epoch.
        perplexity = math.exp(loss) if loss < 300 else float('inf')
        print ("global step %d learning rate %.4f step-time %.2f perplexity "
               "%.2f" % (model.global_step.eval(), model.learning_rate.eval(),
                         step_time, perplexity))
        # Decrease learning rate if no improvement was seen over last 3 times.
        if len(previous_losses) > 2 and loss > max(previous_losses[-3:]):
            sess.run(model.learning_rate_decay_op)
        previous_losses.append(loss)
        # Save checkpoint and zero timer and loss.
        checkpoint_path = os.path.join(FLAGS.train_dir, "translate.ckpt")
        model.saver.save(sess, checkpoint_path, global_step=model.global_step)
        step_time, loss = 0.0, 0.0
```

图 5-7

操作每进行 X 步,都会使用 `previous_losses.append(loss)` 来保存进度,这一

点很重要，因为需要将当前批的损失值与之前的损失值进行比较。当损失值开始增大时，将使用 `sess.run(model.learning_rate_decay_op)` 减小学习率，并在 `dev_set` 上评估损失值，这与之前章节介绍的使用验证集相同，代码如图 5-8 所示。

```python
# Run evals on development set and print their perplexity.
for bucket_id in xrange(len(_buckets)):
    if len(dev_set[bucket_id]) == 0:
        print("  eval: empty bucket %d" % (bucket_id))
        continue
    encoder_inputs, decoder_inputs, target_weights = model.get_batch(
        dev_set, bucket_id)
    _, eval_loss, _ = model.step(sess, encoder_inputs, decoder_inputs,
                                 target_weights, bucket_id, True)
    eval_ppx = math.exp(eval_loss) if eval_loss < 300 else float('inf')
    print("  eval: bucket %d perplexity %.2f" % (bucket_id, eval_ppx))
sys.stdout.flush()
```

图 5-8

运行图 5-8 所示的代码，将得到以下输出。

```
put_count=2530 evicted_count=2000 eviction_rate=0.790514 and
 unsatisfied allocation rate=0
global step 200 learning rate 0.5000 step-time 0.94 perplexity
 1625.06
  eval: bucket 0 perplexity 700.69
  eval: bucket 1 perplexity 433.03
  eval: bucket 2 perplexity 401.39
  eval: bucket 3 perplexity 312.34
global step 400 learning rate 0.5000 step-time 0.91 perplexity
 384.01
  eval: bucket 0 perplexity 124.89
  eval: bucket 1 perplexity 176.36
  eval: bucket 2 perplexity 207.67
  eval: bucket 3 perplexity 239.19
global step 600 learning rate 0.5000 step-time 0.87 perplexity
 266.71
  eval: bucket 0 perplexity 75.80
  eval: bucket 1 perplexity 135.31
  eval: bucket 2 perplexity 167.71
  eval: bucket 3 perplexity 188.42
global step 800 learning rate 0.5000 step-time 0.92 perplexity
 235.76
  eval: bucket 0 perplexity 107.33
  eval: bucket 1 perplexity 159.91
  eval: bucket 2 perplexity 177.93
```

```
eval: bucket 3 perplexity 263.84
```

我们每隔 200 步将看到一条输出内容。这只是所使用的设置之一,我们在文件顶部还定义了以下设置。

```
tf.app.flags.DEFINE_float("learning_rate"", 0.5, ""Learning
                          rate."")
tf.app.flags.DEFINE_float("learning_rate_decay_factor"", 0.99,
                "Learning rate decays by this much."")
tf.app.flags.DEFINE_float("max_gradient_norm"", 5.0,
                "Clip gradients to this norm."")
tf.app.flags.DEFINE_integer("batch_size"", 64,
                "Batch size to use during training."")
tf.app.flags.DEFINE_integer("en_vocab_size"", 40000, ""Size
....."")
tf.app.flags.DEFINE_integer("fr_vocab_size"", 40000, ""Size
                    of...."")
tf.app.flags.DEFINE_integer("size"", 1024, ""Size of each
                    model..."")
tf.app.flags.DEFINE_integer("num_layers"", 3, ""#layers in th
        emodel."")tf.app.flags.DEFINE_string("train_dir"",
 os.path.realpath(''../../datasets/WMT''), ""Training
directory."")
tf.app.flags.DEFINE_integer("max_train_data_size"", 0,
                "Limit size of training data "")
tf.app.flags.DEFINE_integer("steps_per_checkpoint"", 200,
                "Training steps to do per
                    checkpoint."")
```

在构建模型对象时,这些设置的大部分内容将被用到。设置的最后一段就是模型本身,来看一下最后一段。我们回到项目里 3 个文件中的第 3 个即最后一个文件:seq2seq_model.py。

创建了 TensorFlow 会话,就可以在训练过程开始时创建模型了。我们定义的大多数参数用于初始化下面的模型。

```
model = seq2seq_model.Seq2SeqModel(
  FLAGS.en_vocab_size, FLAGS.fr_vocab_size, _buckets,
  FLAGS.size, FLAGS.num_layers, FLAGS.max_gradient_norm,
   FLAGS.batch_size,
  FLAGS.learning_rate, FLAGS.learning_rate_decay_factor,
  forward_only=forward_only)
```

然而,初始化操作是在 seq2seq_model.py 中完成的,因此我们直接跳到此处。

你会发现该模型比较庞大，这就是我们没有逐行解释而是逐块解释的原因。

第一部分是模型的初始化，如图 5-9 和图 5-10 所示。

```python
class Seq2SeqModel(object):
  def __init__(self, source_vocab_size, target_vocab_size, buckets, size,
               num_layers, max_gradient_norm, batch_size, learning_rate,
               learning_rate_decay_factor, use_lstm=False,
               num_samples=512, forward_only=False):

    self.source_vocab_size = source_vocab_size
    self.target_vocab_size = target_vocab_size
    self.buckets = buckets
    self.batch_size = batch_size
    self.learning_rate = tf.Variable(float(learning_rate), trainable=False)
    self.learning_rate_decay_op = self.learning_rate.assign(
        self.learning_rate * learning_rate_decay_factor)
    self.global_step = tf.Variable(0, trainable=False)

    # If we use sampled softmax, we need an output projection.
    output_projection = None
    softmax_loss_function = None
    # Sampled softmax only makes sense if we sample less than vocabulary size.
    if num_samples > 0 and num_samples < self.target_vocab_size:
      with tf.device("/cpu:0"):
        w = tf.get_variable("proj_w", [size, self.target_vocab_size])
        w_t = tf.transpose(w)
        b = tf.get_variable("proj_b", [self.target_vocab_size])
      output_projection = (w, b)

      def sampled_loss(inputs, labels):
        with tf.device("/cpu:0"):
          labels = tf.reshape(labels, [-1, 1])
          return tf.nn.sampled_softmax_loss(w_t, b, inputs, labels, num_samples,
                                            self.target_vocab_size)
      softmax_loss_function = sampled_loss

    # Create the internal multi-layer cell for our RNN.
    single_cell = tf.nn.rnn_cell.GRUCell(size)
    if use_lstm:
      single_cell = tf.nn.rnn_cell.BasicLSTMCell(size)
    cell = single_cell
    if num_layers > 1:
      cell = tf.nn.rnn_cell.MultiRNNCell([single_cell] * num_layers)

    # The seq2seq function: we use embedding for the input and attention.
    def seq2seq_f(encoder_inputs, decoder_inputs, do_decode):
      return tf.nn.seq2seq.embedding_attention_seq2seq(
          encoder_inputs, decoder_inputs, cell,
          num_encoder_symbols=source_vocab_size,
          num_decoder_symbols=target_vocab_size,
          embedding_size=size,
          output_projection=output_projection,
          feed_previous=do_decode)
```

图 5-9

```python
# Feeds for inputs.
self.encoder_inputs = []
self.decoder_inputs = []
self.target_weights = []
for i in xrange(buckets[-1][0]):  # Last bucket is the biggest one.
    self.encoder_inputs.append(tf.placeholder(tf.int32, shape=[None],
                                              name="encoder{0}".format(i)))
for i in xrange(buckets[-1][1] + 1):
    self.decoder_inputs.append(tf.placeholder(tf.int32, shape=[None],
                                              name="decoder{0}".format(i)))
    self.target_weights.append(tf.placeholder(tf.float32, shape=[None],
                                              name="weight{0}".format(i)))

# Our targets are decoder inputs shifted by one.
targets = [self.decoder_inputs[i + 1]
           for i in xrange(len(self.decoder_inputs) - 1)]

# Training outputs and losses.
if forward_only:
    self.outputs, self.losses = tf.nn.seq2seq.model_with_buckets(
        self.encoder_inputs, self.decoder_inputs, targets,
        self.target_weights, buckets, lambda x, y: seq2seq_f(x, y, True),
        softmax_loss_function=softmax_loss_function)
    # If we use output projection, we need to project outputs for decoding.
    if output_projection is not None:
        for b in xrange(len(buckets)):
            self.outputs[b] = [
                tf.matmul(output, output_projection[0]) + output_projection[1]
                for output in self.outputs[b]
            ]
else:
    self.outputs, self.losses = tf.nn.seq2seq.model_with_buckets(
        self.encoder_inputs, self.decoder_inputs, targets,
        self.target_weights, buckets,
        lambda x, y: seq2seq_f(x, y, False),
        softmax_loss_function=softmax_loss_function)

# Gradients and SGD update operation for training the model.
params = tf.trainable_variables()
if not forward_only:
    self.gradient_norms = []
    self.updates = []
    opt = tf.train.GradientDescentOptimizer(self.learning_rate)
    for b in xrange(len(buckets)):
        gradients = tf.gradients(self.losses[b], params)
        clipped_gradients, norm = tf.clip_by_global_norm(gradients,
                                                         max_gradient_norm)
        self.gradient_norms.append(norm)
        self.updates.append(opt.apply_gradients(
            zip(clipped_gradients, params), global_step=self.global_step))

self.saver = tf.train.Saver(tf.all_variables())
```

图 5-10

模型从初始化开始，在初始化阶段设置了模型所需要的参数。我们将跳过这些参数的设置，因为在训练之前，我们已经将值传递到模型构造语句中，并初始化了这些参数，它们最终通过 self.xyz 赋值传递到内部变量中。

传入每个模型层的大小（size=1024）和层数（3）对于构造权重和偏差（proj_w 和 proj_b）非常重要。权重是 $A \times B$，其中 A 是层的大小，B 是目标语言的词汇表大小。偏差是根据目标词汇表的大小来传递的。

最后，使用来自元组 output_project 中的权重和偏差（output_projection = (w, b)），以及转置后的权重和偏差来构成 softmax_loss_function，我们将会反复使用它来衡量性能。

模型的下一部分是 step 函数，如图 5-11 所示。step 函数的前半部分仅是错误检查，所以直接跳过它。最有趣的是利用随机梯度下降法构造输出反馈。

```python
def step(self, session, encoder_inputs, decoder_inputs, target_weights,
        bucket_id, forward_only):
    encoder_size, decoder_size = self.buckets[bucket_id]
    if len(encoder_inputs) != encoder_size:
        raise ValueError("Encoder length must be equal to the one in bucket,"
                         " %d != %d." % (len(encoder_inputs), encoder_size))
    if len(decoder_inputs) != decoder_size:
        raise ValueError("Decoder length must be equal to the one in bucket,"
                         " %d != %d." % (len(decoder_inputs), decoder_size))
    if len(target_weights) != decoder_size:
        raise ValueError("Weights length must be equal to the one in bucket,"
                         " %d != %d." % (len(target_weights), decoder_size))

    # Input feed: encoder inputs, decoder inputs, target_weights, as provided.
    input_feed = {}
    for l in xrange(encoder_size):
        input_feed[self.encoder_inputs[l].name] = encoder_inputs[l]
    for l in xrange(decoder_size):
        input_feed[self.decoder_inputs[l].name] = decoder_inputs[l]
        input_feed[self.target_weights[l].name] = target_weights[l]

    # Since our targets are decoder inputs shifted by one, we need one more.
    last_target = self.decoder_inputs[decoder_size].name
    input_feed[last_target] = np.zeros([self.batch_size], dtype=np.int32)

    # Output feed: depends on whether we do a backward step or not.
    if not forward_only:
        output_feed = [self.updates[bucket_id],  # Update Op that does SGD.
                       self.gradient_norms[bucket_id],  # Gradient norm.
                       self.losses[bucket_id]]  # Loss for this batch.
    else:
        output_feed = [self.losses[bucket_id]]  # Loss for this batch.
        for l in xrange(decoder_size):  # Output logits.
            output_feed.append(self.outputs[bucket_id][l])

    outputs = session.run(output_feed, input_feed)
    if not forward_only:
        return outputs[1], outputs[2], None  # Gradient norm, loss, no outputs.
    else:
        return None, outputs[0], outputs[1:]  # No gradient norm, loss, outputs.
```

图 5-11

模型的最后一部分是 get_batch 函数，如图 5-12 所示。可以通过内部的注释来了解这个函数的各个部分。

```python
def get_batch(self, data, bucket_id):

    encoder_size, decoder_size = self.buckets[bucket_id]
    encoder_inputs, decoder_inputs = [], []

    # Get a random batch of encoder and decoder inputs from data,
    # pad them if needed, reverse encoder inputs and add GO to decoder.
    for _ in xrange(self.batch_size):
        encoder_input, decoder_input = random.choice(data[bucket_id])

        # Encoder inputs are padded and then reversed.
        encoder_pad = [data_utils.PAD_ID] * (encoder_size - len(encoder_input))
        encoder_inputs.append(list(reversed(encoder_input + encoder_pad)))

        # Decoder inputs get an extra "GO" symbol, and are padded then.
        decoder_pad_size = decoder_size - len(decoder_input) - 1
        decoder_inputs.append([data_utils.GO_ID] + decoder_input +
                              [data_utils.PAD_ID] * decoder_pad_size)

    # Now we create batch-major vectors from the data selected above.
    batch_encoder_inputs, batch_decoder_inputs, batch_weights = [], [], []

    # Batch encoder inputs are just re-indexed encoder_inputs.
    for length_idx in xrange(encoder_size):
        batch_encoder_inputs.append(
            np.array([encoder_inputs[batch_idx][length_idx]
                      for batch_idx in xrange(self.batch_size)], dtype=np.int32))

    # Batch decoder inputs are re-indexed decoder_inputs, we create weights.
    for length_idx in xrange(decoder_size):
        batch_decoder_inputs.append(
            np.array([decoder_inputs[batch_idx][length_idx]
                      for batch_idx in xrange(self.batch_size)], dtype=np.int32))

        # Create target_weights to be 0 for targets that are padding.
        batch_weight = np.ones(self.batch_size, dtype=np.float32)
        for batch_idx in xrange(self.batch_size):
            # We set weight to 0 if the corresponding target is a PAD symbol.
            # The corresponding target is decoder_input shifted by 1 forward.
            if length_idx < decoder_size - 1:
                target = decoder_inputs[batch_idx][length_idx + 1]
            if length_idx == decoder_size - 1 or target == data_utils.PAD_ID:
                batch_weight[batch_idx] = 0.0
        batch_weights.append(batch_weight)
    return batch_encoder_inputs, batch_decoder_inputs, batch_weights
```

图 5-12

当运行图 5-12 所示代码时，我们可以得到一个完美的训练运行结果，如下所示。

```
global step 200 learning rate 0.5000 step-time 0.94 perplexity
 1625.06
  eval: bucket 0 perplexity 700.69
  eval: bucket 1 perplexity 433.03
  eval: bucket 2 perplexity 401.39
  eval: bucket 3 perplexity 312.34
  ...
```

你可能会发现其中存在这样一些方式，例如，在损失值持续增长之后我们降低了学习率。无论哪种方式，我们都将持续在开发集上进行测试，直到准确率有所提高。

5.4 总结

本章介绍了序列到序列模型，并使用一系列已知的句子-句子翻译作为训练集，来编写了一个语言翻译器。本章还引入了 RNNs 作为工作的基础，并初步接触了大数据，因为训练时使用了一个 20 GB 的训练数据集。

接下来，我们将开始应用表格数据，对经济和金融数据进行预测。我们将使用之前工作的一部分内容，这样就可以运行基本的准备工作，即我们已经编写的用于下载和准备训练数据的初始管道工作。后续我们将专注于时间序列问题，它与目前所做的图像和文本工作大不相同。

第 6 章
探索文本含义

到目前为止，我们主要使用 TensorFlow 来进行图像处理，并在某些时候将其用于文本序列处理。在本章中，我们将回顾手写字体识别并寻找文本中的含义。这通常是**自然语言处理**（Natural Language Processing，NLP）领域的一部分。该领域的一些应用如下所示。

- 情感分析——它从文本中提取普通的情绪类别，而无须提取句子的主题或动作。
- 实体抽取——它会提取主题，例如，从一段文本中提取人物、地点和事件。
- 关键字提取——它会从一段文本中提取关键字。
- 词关系提取——它不仅会提取实体，还会提取相关的动作以及词性。

以上仅仅涉及了 NLP 的表皮，NLP 还包含其他技术，以及这些技术的一系列复杂的衍生问题。这看起来似乎有点学术，但从上面介绍的 4 种技术能实现的内容来考虑，就不会这样认为了。以下是一些应用示例。

- 阅读新闻并理解新闻的主题（主体、公司、位置等）。
- 阅读前面的新闻并理解它们所包含的情绪（快乐、悲伤、愤怒等）。
- 分析产品评论并理解用户对该产品的情绪（满意、失望等）。
- 编写一个机器人，响应自然语言形式的用户聊天命令。

与之前的机器学习探索工作相同，我们付出了相当多的努力来进行设置。在本例中，我们将花一些时间来编写脚本，以真正地实现从兴趣源头获取文本。

6.1 额外设置

为了包含文本处理所需要的库,需要进行额外的设置。如下面所示。

1. 首先是 Bazel。在 Ubuntu 上,根据官方教程安装 Bazel。在 macOS 上,可以使用 HomeBrew 来安装 Bazel,如下所示。

```
$ brew install bazel
```

2. 然后,安装 swig,它允许我们封装 C/C++ 函数,以实现在 Python 中对这些函数进行调用。在 Ubuntu 上,可以使用以下命令安装。

```
$ sudo apt-get install swig
```

在 macOS 上,使用 brew 来安装,如下所示。

```
$ brew install swig
```

3. 接下来安装协议缓冲区支持,它允许我们以一种比 XML 更有效的方式来存储和检索序列化数据。这里需要使用 3.3.0 版本,可以通过以下命令安装。

```
$ pip install -U protobuf==3.3.0
```

4. 我们的文本分类将以树的形式来表示,因此需要一个工具库来实现在命令行上展示树。我们通过如下方式安装该工具库。

```
$ pip install asciitree
```

5. 最后,我们还需要一个科学计算库 NumPy。你可以参考图像分类的相关内容,或者用下面的命令安装 NumPy。

```
$ pip install numpy autograd
```

完成以上步骤之后,就可以安装 SyntaxNet 了,它为我们的 NLP 做了大量的工作。SyntaxNet 是 TensorFlow 的一个开源框架,它提供了基本的功能。谷歌利用英语训练了一个 SyntaxNet 模型,并将其命名为 **Parsey McParseface**,它将包含在我们的安装软件中。这样,我们既可以用英语训练自己的较为具体的模型,也可以使用其他语言来训练模型。

同往常一样，训练数据将会带来一个挑战。我们首先使用预训练的英文模型 Parsey McParseface。

获取相应的包并对其进行配置，相关命令如下所示。

```
$ git clone --recursive GitHub官网 tensorflow/models.git
$ cd models/research/syntaxnet/tensorflow
$ ./configure
```

最后，按以下方式对系统进行测试。

```
$ cd ...
$ bazel test ...
```

测试需要一段时间，你需要耐心等待。如果你严格执行了所有指令，那么所有的测试都将会通过。但是在我们的计算机上可能会出现下面的一些错误。

- bazel 无法下载包。可以尝试使用以下命令重新下载，然后再次运行测试命令。

```
$ bazel clean --expunge
```

- 如果测试失败，那么将以下代码添加到 home 目录下的 .bazelrc 文件中，以便了解更多的错误信息来进行调试。

```
test --test_output=errors
```

- 如果遇到错误 "Tensor already registered"，那么需要按照 GitHub 上的解决方案来处理。

现在来做一个更普通的测试。提供一个英语句子，然后查看它是如何被解析的。

```
$ echo 'Faaris likes to feed the kittens.' | bash ./syntaxnet/demo.sh
```

通过 echo 语句输入一个句子，然后利用管道将它输入 SyntaxNet 演示脚本，该脚本会从控制台接收标准输入。需要注意的是，为了使示例更加有趣，这里使用了一个不寻常的名字，例如 Faaris。运行命令将生成大量的调试信息，如下所示，这里用...截断了堆栈跟踪信息。

```
I syntaxnet/term_frequency_map.cc:101] Loaded 46 terms from syntaxnet/models/parsey_mcparseface/label-map.
I syntaxnet/embedding_feature_extractor.cc:35] Features: input.digit input.hyphen; input.prefix(length="2") input(1).prefix(length="2")
```

```
input(2).prefix(length="2") input(3).prefix(length="2") input(-
1).prefix(length="2")...
    I syntaxnet/embedding_feature_extractor.cc:36] Embedding names:
other;prefix2;prefix3;suffix2;suffix3;words
    I syntaxnet/embedding_feature_extractor.cc:37] Embedding dims:
8;16;16;16;16;64
    I syntaxnet/term_frequency_map.cc:101] Loaded 46 terms from
syntaxnet/models/parsey_mcparseface/label-map.
    I syntaxnet/embedding_feature_extractor.cc:35] Features:
stack.child(1).label stack.child(1).sibling(-1).label stack.child(-
1)...
    I syntaxnet/embedding_feature_extractor.cc:36] Embedding names:
labels;tags;words
    I syntaxnet/embedding_feature_extractor.cc:37] Embedding dims:
32;32;64
    I syntaxnet/term_frequency_map.cc:101] Loaded 49 terms from
syntaxnet/models/parsey_mcparseface/tag-map.
    I syntaxnet/term_frequency_map.cc:101] Loaded 64036 terms from
syntaxnet/models/parsey_mcparseface/word-map.
    I syntaxnet/term_frequency_map.cc:101] Loaded 64036 terms from
syntaxnet/models/parsey_mcparseface/word-map.
    I syntaxnet/term_frequency_map.cc:101] Loaded 49 terms from
syntaxnet/models/parsey_mcparseface/tag-map.
    INFO:tensorflow:Building training network with parameters:
feature_sizes: [12 20 20] domain_sizes: [   49    51 64038]
    INFO:tensorflow:Building training network with parameters:
feature_sizes: [2 8 8 8 8 8] domain_sizes: [    5 10665 10665    8970
 8970 64038]
    I syntaxnet/term_frequency_map.cc:101] Loaded 46 terms from
syntaxnet/models/parsey_mcparseface/label-map.
    I syntaxnet/embedding_feature_extractor.cc:35] Features:
stack.child(1).label stack.child(1).sibling(-1).label stack.child(-
1)...
    I syntaxnet/embedding_feature_extractor.cc:36] Embedding names:
labels;tags;words
    I syntaxnet/embedding_feature_extractor.cc:37] Embedding dims:
32;32;64
    I syntaxnet/term_frequency_map.cc:101] Loaded 49 terms from
syntaxnet/models/parsey_mcparseface/tag-map.
    I syntaxnet/term_frequency_map.cc:101] Loaded 64036 terms from
syntaxnet/models/parsey_mcparseface/word-map.
    I syntaxnet/term_frequency_map.cc:101] Loaded 49 terms from
syntaxnet/models/parsey_mcparseface/tag-map.
```

```
    I syntaxnet/term_frequency_map.cc:101] Loaded 46 terms from
syntaxnet/models/parsey_mcparseface/label-map.
    I syntaxnet/embedding_feature_extractor.cc:35] Features: input.digit
input.hyphen; input.prefix(length="2") input(1).prefix(length="2")
input(2).prefix(length="2") input(3).prefix(length="2") input(-
1).prefix(length="2")...
    I syntaxnet/embedding_feature_extractor.cc:36] Embedding names:
other;prefix2;prefix3;suffix2;suffix3;words
    I syntaxnet/embedding_feature_extractor.cc:37] Embedding dims:
8;16;16;16;16;64
    I syntaxnet/term_frequency_map.cc:101] Loaded 64036 terms from
syntaxnet/models/parsey_mcparseface/word-map.
    INFO:tensorflow:Processed 1 documents
    INFO:tensorflow:Total processed documents: 1
    INFO:tensorflow:num correct tokens: 0
    INFO:tensorflow:total tokens: 7
    INFO:tensorflow:Seconds elapsed in evaluation: 0.12, eval metric:
0.00%
    INFO:tensorflow:Processed 1 documents
    INFO:tensorflow:Total processed documents: 1
    INFO:tensorflow:num correct tokens: 1
    INFO:tensorflow:total tokens: 6
    INFO:tensorflow:Seconds elapsed in evaluation: 0.47, eval metric:
16.67%
    INFO:tensorflow:Read 1 documents
    Input: Faaris likes to feed the kittens .
    Parse:
    likes VBZ ROOT
     +-- Faaris NNP nsubj
     +-- feed VB xcomp
     |   +-- to TO aux
     |   +-- kittens NNS dobj
     |       +-- the DT det
     +-- . . punct
```

以 "Input:" 起始的这一部分（最后一部分）是最有趣的部分。我们可以以编程的方式使用这一部分。请注意句子是如何分解成言语和实体-动作-对象对的一部分的。可以看到，一些单词的名称为：nsubj、xcomp、aux、dobj、det 和 punct。其中一些名称是显而易见的，而其他的则并非如此。如果要深入研究，建议仔细学习斯坦福依赖层次结构（Stanford dependenay hierarchy）。

在继续下一步之前,我们用另一个句子进行训练。

```
Input: Stop speaking so loudly and be quiet !
Parse:
Stop VB ROOT
+-- speaking VBG xcomp
|    +-- loudly RB advmod
|        +-- so RB advmod
|        +-- and CC cc
|        +-- quiet JJ conj
|            +-- be VB cop
+-- ! . punct
```

这再次验证了该模型在剖析短语时表现良好。读者可以尝试一些自己的句子。

接下来,我们开始真正训练一个模型。训练 SyntaxNet 的过程是相当琐碎的,因为它是一个编译系统。到目前为止,我们已经通过标准输入(STDIO)导入了数据,也可以通过管道输入文本语料库。我们使用协议缓冲器库来编辑源文件——syntaxnet/ models/parsey_mcparseface/context.pbtxt。

另外,我们将把训练源更改为其他训练源或者我们自己的训练源,如下面代码所示。

```
input {
 name: 'wsj-data'
 record_format: 'conll-sentence'
 Part {
   file_pattern: './wsj.conll'
  }
}
input {
 name: 'wsj-data-tagged'
 record_format: 'conll-sentence'
 Part {
   file_pattern: './wsj-tagged.conll'
  }
}
```

这就是我们训练数据集的方法。然而,要想做得比天然训练模型 Parsey McParseface 更好,那你需要找到更好的模型,这会很有挑战性。我们将使用一个新的模型——**卷积神经网络(CNN)** 来对一个有趣的数据集进行训练,以用来处理文本。

我们使用的是由康奈尔大学计算机科学系汇编的电影评论数据。

首先，下载并处理电影评论数据集，然后利用它训练模型，最后在此基础上进行评估。所有的代码由 Denny Britz 在 GitHub 上维护。

这段代码受到了 Yoon Kim 关于"CNNs for sentence classification"这个主题的论文启发，并由谷歌的 Denny Britz 实现和维护。现在，我们将阅读此代码，查看 Danny Britz 是如何实现该网络的。

首先，利用常用的数据辅助函数来下载并准备特定的数据集，如图 6-1 所示。

```python
import tensorflow as tf
import numpy as np
import os
import time
import datetime
import data_helpers
from text_cnn import TextCNN

# Parameters
# ==================================================

# Model Hyperparameters
tf.flags.DEFINE_integer("embedding_dim", 128, "Dimensionality of character embedding (default: 128)")
tf.flags.DEFINE_string("filter_sizes", "3,4,5", "Comma-separated filter sizes (default: '3,4,5')")
tf.flags.DEFINE_integer("num_filters", 128, "Number of filters per filter size (default: 128)")
tf.flags.DEFINE_float("dropout_keep_prob", 0.5, "Dropout keep probability (default: 0.5)")
tf.flags.DEFINE_float("l2_reg_lambda", 0.0, "L2 regularizaion lambda (default: 0.0)")

# Training parameters
tf.flags.DEFINE_integer("batch_size", 64, "Batch Size (default: 64)")
tf.flags.DEFINE_integer("num_epochs", 200, "Number of training epochs (default: 200)")
tf.flags.DEFINE_integer("evaluate_every", 100, "Evaluate model on dev set after this many steps (default: 100)")
tf.flags.DEFINE_integer("checkpoint_every", 100, "Save model after this many steps (default: 100)")
# Misc Parameters
tf.flags.DEFINE_boolean("allow_soft_placement", True, "Allow device soft device placement")
tf.flags.DEFINE_boolean("log_device_placement", False, "Log placement of ops on devices")

FLAGS = tf.flags.FLAGS
FLAGS._parse_flags()
print("\nParameters:")
for attr, value in sorted(FLAGS.__flags.items()):
    print("{}={}".format(attr.upper(), value))
print("")
```

图 6-1

接着，定义参数。这些参数定义了每次循环中处理数据的大小，以及是部分运行代码还是完全运行代码。我们还将定义评估进展的频率（此处为 100 步），以及为模型保存检查点的频率（以允许评估和重新延续）。加载和准备数据集的代码如图 6-2 所示。

```python
# Data Preparation
# ==================================================

# Load data
print("Loading data...")
x_text, y = data_helpers.load_data_and_labels(FLAGS.positive_data_file, FLAGS.negative_data_file)

# Build vocabulary
max_document_length = max([len(x.split(" ")) for x in x_text])
vocab_processor = learn.preprocessing.VocabularyProcessor(max_document_length)
x = np.array(list(vocab_processor.fit_transform(x_text)))

# Randomly shuffle data
np.random.seed(10)
shuffle_indices = np.random.permutation(np.arange(len(y)))
x_shuffled = x[shuffle_indices]
y_shuffled = y[shuffle_indices]

# Split train/test set
# TODO: This is very crude, should use cross-validation
dev_sample_index = -1 * int(FLAGS.dev_sample_percentage * float(len(y)))
x_train, x_dev = x_shuffled[:dev_sample_index], x_shuffled[dev_sample_index:]
y_train, y_dev = y_shuffled[:dev_sample_index], y_shuffled[dev_sample_index:]
print("Vocabulary Size: {:d}".format(len(vocab_processor.vocabulary_)))
print("Train/Dev split: {:d}/{:d}".format(len(y_train), len(y_dev)))
```

图 6-2

代码中的训练部分如图 6-3 所示。

```python
# Training
# ==================================================
with tf.Graph().as_default():
    session_conf = tf.ConfigProto(
        allow_soft_placement=FLAGS.allow_soft_placement,
        log_device_placement=FLAGS.log_device_placement)
    sess = tf.Session(config=session_conf)
    with sess.as_default():
        cnn = TextCNN(
            sequence_length=x_train.shape[1],
            num_classes=2,
            vocab_size=len(vocabulary),
            embedding_size=FLAGS.embedding_dim,
            filter_sizes=list(map(int, FLAGS.filter_sizes.split(","))),
            num_filters=FLAGS.num_filters,
            l2_reg_lambda=FLAGS.l2_reg_lambda)

        # Define Training procedure
        global_step = tf.Variable(0, name="global_step", trainable=False)
        optimizer = tf.train.AdamOptimizer(1e-3)
        grads_and_vars = optimizer.compute_gradients(cnn.loss)
        train_op = optimizer.apply_gradients(grads_and_vars, global_step=global_step)

        # Keep track of gradient values and sparsity (optional)
        grad_summaries = []
        for g, v in grads_and_vars:
            if g is not None:
                grad_hist_summary = tf.histogram_summary("{}/grad/hist".format(v.name), g)
                sparsity_summary = tf.scalar_summary("{}/grad/sparsity".format(v.name), tf.nn.zero_fraction(g))
                grad_summaries.append(grad_hist_summary)
                grad_summaries.append(sparsity_summary)
        grad_summaries_merged = tf.merge_summary(grad_summaries)

        # Output directory for models and summaries
        timestamp = str(int(time.time()))
        out_dir = os.path.abspath(os.path.join(os.path.curdir, "runs", timestamp))
        print("Writing to {}\n".format(out_dir))
```

图 6-3

图 6-3 展示了 CNN 的实例化——一个自然语言处理下的 CNN，其中包含一些前面定义的参数。除此之外，我们还编写了启用 TensorBoard 可视化的代码。

图 6-4 展示了我们为 TensorBoard 捕捉的更多指标——损失值、训练的准确率，以及评估集。

```
# Summaries for loss and accuracy
loss_summary = tf.scalar_summary("loss", cnn.loss)
acc_summary = tf.scalar_summary("accuracy", cnn.accuracy)

# Train Summaries
train_summary_op = tf.merge_summary([loss_summary, acc_summary, grad_summaries_merged])
train_summary_dir = os.path.join(out_dir, "summaries", "train")
train_summary_writer = tf.train.SummaryWriter(train_summary_dir, sess.graph_def)

# Dev summaries
dev_summary_op = tf.merge_summary([loss_summary, acc_summary])
dev_summary_dir = os.path.join(out_dir, "summaries", "dev")
dev_summary_writer = tf.train.SummaryWriter(dev_summary_dir, sess.graph_def)

# Checkpoint directory. Tensorflow assumes this directory already exists so we need to create it
checkpoint_dir = os.path.abspath(os.path.join(out_dir, "checkpoints"))
checkpoint_prefix = os.path.join(checkpoint_dir, "model")
if not os.path.exists(checkpoint_dir):
    os.makedirs(checkpoint_dir)
saver = tf.train.Saver(tf.all_variables())
```

图 6-4

接下来我们将定义训练和评估方法，这些方法与前文提及的图像处理方法类似。我们将收到一组训练数据和标签，并将它们存储在一个字典中。然后，对字典运行 TensorFlow 会话，并捕获返回的性能指标。

我们在代码开始处定义训练方法和评估方法，然后批量地循环训练数据，并将这些方法应用于每一批数据。在选择的时间间隔内，我们还将为可选的评估保存检查点，实现代码如图 6-5 所示。

```
def train_step(x_batch, y_batch):
    """
    A single training step
    """
    feed_dict = {
      cnn.input_x: x_batch,
      cnn.input_y: y_batch,
      cnn.dropout_keep_prob: FLAGS.dropout_keep_prob
    }
    _, step, summaries, loss, accuracy = sess.run(
        [train_op, global_step, train_summary_op, cnn.loss, cnn.accuracy],
```

图 6-5

```python
          feed_dict)
    time_str = datetime.datetime.now().isoformat()
    print("{}: step {}, loss {:g}, acc {:g}".format(time_str, step, loss, accuracy))
    train_summary_writer.add_summary(summaries, step)

def dev_step(x_batch, y_batch, writer=None):
    """
    Evaluates model on a dev set
    """
    feed_dict = {
      cnn.input_x: x_batch,
      cnn.input_y: y_batch,
      cnn.dropout_keep_prob: 1.0
    }
    step, summaries, loss, accuracy = sess.run(
        [global_step, dev_summary_op, cnn.loss, cnn.accuracy],
        feed_dict)
    time_str = datetime.datetime.now().isoformat()
    print("{}: step {}, loss {:g}, acc {:g}".format(time_str, step, loss, accuracy))
    if writer:
        writer.add_summary(summaries, step)

# Generate batches
batches = data_helpers.batch_iter(
    list(zip(x_train, y_train)), FLAGS.batch_size, FLAGS.num_epochs)
# Training loop. For each batch...
for batch in batches:
    x_batch, y_batch = zip(*batch)
    train_step(x_batch, y_batch)
    current_step = tf.train.global_step(sess, global_step)
    if current_step % FLAGS.evaluate_every == 0:
        print("\nEvaluation:")
        dev_step(x_dev, y_dev, writer=dev_summary_writer)
        print("")
    if current_step % FLAGS.checkpoint_every == 0:
        path = saver.save(sess, checkpoint_prefix, global_step=current_step)
        print("Saved model checkpoint to {}\n".format(path))
```

图 6-5（续）

在一台仅支持 CPU 的机器上运行以上代码，经过一小时的训练，最终得到一个训练好的模型，该模型将被存储为一个检查点文件。把该文件输入评估程序，如图 6-6 所示。

```python
#! /usr/bin/env python

import tensorflow as tf
import numpy as np
import os
import time
import datetime
import data_helpers
from text_cnn import TextCNN

# Parameters
# ==================================================

# Eval Parameters
tf.flags.DEFINE_integer("batch_size", 64, "Batch Size (default: 64)")
tf.flags.DEFINE_string("checkpoint_dir", "", "Checkpoint directory from training run")

# Misc Parameters
tf.flags.DEFINE_boolean("allow_soft_placement", True, "Allow device soft device placement")
tf.flags.DEFINE_boolean("log_device_placement", False, "Log placement of ops on devices")


FLAGS = tf.flags.FLAGS
FLAGS._parse_flags()
print("\nParameters:")
for attr, value in sorted(FLAGS.__flags.items()):
    print("{}={}".format(attr.upper(), value))
print("")

# Load data. Load your own data here
print("Loading data...")
x_test, y_test, vocabulary, vocabulary_inv = data_helpers.load_data()
y_test = np.argmax(y_test, axis=1)
print("Vocabulary size: {:d}".format(len(vocabulary)))
print("Test set size {:d}".format(len(y_test)))

print("\nEvaluating...\n")
```

图 6-6

评估程序只是一个使用示例,从图 6-6 所示的评估程序可以看出,我们也会将检查点目录作为输入,并加载一些测试数据。但是,你应该使用自己的数据。

接下来,我们看一下图 6-7 所示的代码。

```python
print("\nEvaluating...\n")

# Evaluation
# ==================================================
checkpoint_file = tf.train.latest_checkpoint(FLAGS.checkpoint_dir)
graph = tf.Graph()
with graph.as_default():
    session_conf = tf.ConfigProto(
      allow_soft_placement=FLAGS.allow_soft_placement,
      log_device_placement=FLAGS.log_device_placement)
    sess = tf.Session(config=session_conf)
    with sess.as_default():
        # Load the saved meta graph and restore variables
        saver = tf.train.import_meta_graph("{}.meta".format(checkpoint_file))
        saver.restore(sess, checkpoint_file)

        # Get the placeholders from the graph by name
        input_x = graph.get_operation_by_name("input_x").outputs[0]
        # input_y = graph.get_operation_by_name("input_y").outputs[0]
        dropout_keep_prob = graph.get_operation_by_name("dropout_keep_prob").outputs[0]

        # Tensors we want to evaluate
        predictions = graph.get_operation_by_name("output/predictions").outputs[0]

        # Generate batches for one epoch
        batches = data_helpers.batch_iter(x_test, FLAGS.batch_size, 1, shuffle=False)

        # Collect the predictions here
        all_predictions = []

        for x_test_batch in batches:
            batch_predictions = sess.run(predictions, {input_x: x_test_batch, dropout_keep_prob: 1.0})
            all_predictions = np.concatenate([all_predictions, batch_predictions])

# Print accuracy
correct_predictions = float(sum(all_predictions == y_test))
print("Total number of test examples: {}".format(len(y_test)))
print("Accuracy: {:g}".format(correct_predictions/float(len(y_test))))
```

图 6-7

从检查点文件开始。加载检查点文件并重新创建一个 TensorFlow 会话。这样就可以对刚刚训练过的模型进行评估，该会话可以反复使用。

接下来，分批运行测试数据。在正常情况下，我们不会使用循环或批量，但是因为有大量的测试数据，所以我们通过循环来实现它。

简单地对每组测试数据运行会话，并保留返回的预测值（正面或负面）。以下是一些正面评价数据的样本。

```
insomnia loses points when it surrenders to a formulaic bang-bang ,
shoot-em-up scene at the conclusion . but the performances of pacino
, williams , and swank keep the viewer wide-awake all the way through.
```

```
    what might have been readily dismissed as the tiresome rant of an
aging filmmaker still thumbing his nose at convention takes a
surprising , subtle turn at the midway point .
    at a time when commercialism has squeezed the life out of whatever
idealism american moviemaking ever had , godfrey reggio's career
shines like a lonely beacon .
    an inuit masterpiece that will give you goosebumps as its uncanny
tale of love , communal discord , and justice unfolds .
    this is popcorn movie fun with equal doses of action , cheese , ham
and cheek ( as well as a serious debt to the road warrior ) , but it
feels like unrealized potential
    it's a testament to de niro and director michael caton-jones that by
movie's end , we accept the characters and the film , flaws and all .
    performances are potent , and the women's stories are ably intercut
and involving .
    an enormously entertaining movie , like nothing we've ever seen
before , and yet completely familiar .
    lan yu is a genuine love story , full of traditional layers of
awakening and ripening and separation and recovery .
    your children will be occupied for 72 minutes .
    pull[s] off the rare trick of recreating not only the look of a
certain era , but also the feel .
    twohy's a good yarn-spinner , and ultimately the story compels .
    'tobey maguire is a poster boy for the geek generation . '
    . . . a sweetly affecting story about four sisters who are coping ,
in one way or another , with life's endgame .
    passion , melodrama , sorrow , laugther , and tears cascade over the
screen effortlessly . . .
    road to perdition does display greatness , and it's worth seeing .
but it also comes with the laziness and arrogance of a thing that
already knows it's won .
```

类似地，我们也有负面评价的数据。它们分别保存在 data 文件夹中的 rt-polarity.pos 和 rt-polarity.neg 文件中。

我们使用的网络架构如图 6-8 所示。

它与我们用于图像处理的网络架构非常相似。事实上，整个过程都非常相似。很多技术的美妙之处就在于它们的通用性。

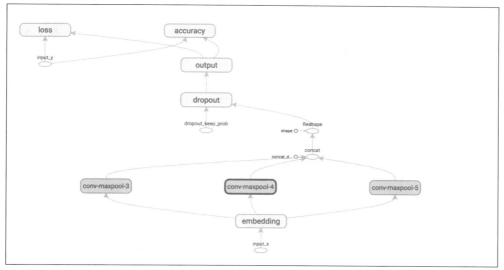

图 6-8

查看训练的输出,如下所示。

```
$ ./train.py
...
2017-06-15T04:42:08.793884: step 30101, loss 0, acc 1
2017-06-15T04:42:08.934489: step 30102, loss 1.54599e-07, acc 1
2017-06-15T04:42:09.082239: step 30103, loss 3.53902e-08, acc 1
2017-06-15T04:42:09.225435: step 30104, loss 0, acc 1
2017-06-15T04:42:09.369348: step 30105, loss 2.04891e-08, acc 1
2017-06-15T04:42:09.520073: step 30106, loss 0.0386909, acc
0.984375
2017-06-15T04:42:09.676975: step 30107, loss 8.00917e-07, acc 1
2017-06-15T04:42:09.821703: step 30108, loss 7.83049e-06, acc 1
...
2017-06-15T04:42:23.220202: step 30199, loss 1.49012e-08, acc 1
2017-06-15T04:42:23.366740: step 30200, loss 5.67226e-05, acc 1
Evaluation:
2017-06-15T04:42:23.781196: step 30200, loss 9.74802, acc 0.721
...
Saved model checkpoint to /Users/saif/Documents/BOOK/cnn-textclassificat
ion-tf/runs/1465950150/checkpoints/model-30200
```

评估步骤如下。

```
$ ./eval.py -eval_train --checkpoint_dir==./runs/1465950150/checkpoints/
Parameters:
```

```
ALLOW_SOFT_PLACEMENT=True
BATCH_SIZE=64
CHECKPOINT_DIR=/Users/saif/Documents/BOOK/cnn-text-classificationtf/
runs/1465950150/checkpoints/
LOG_DEVICE_PLACEMENT=False
Loading data...
Vocabulary size: 18765
Test set size 10662
Evaluating...
Total number of test examples: 10662
Accuracy: 0.973832
```

这在我们的数据集上是相当准确的。下一步工作是将训练的模型应用到日常使用上。首先是获取数据。一些有趣的试验可能是从其他源（例如 IMDB 或亚马逊）获取电影评论数据。因为数据不一定会被标记，所以可以使用正的区分比作为跨站点通用协议的度量标准。

然后，就可以在该领域使用模型了。假设你是一个产品制造商，那么你可以实时追踪各种来源的所有评论，并过滤出非常负面的评论。然后，你的商务代理人可以尝试解决这些问题。

应用有无限种可能性，这里提出一个你可能从事的有趣项目，以将学过的内容结合起来。编写一个 Twitter 流阅读器，它接收每条推文，并提取推文的主题。对于一组特定的主题，例如公司，该阅读器可以评估这条推文是正面的还是负面的；为正负百分比创建运行指标，以不同的时间尺度来评估主题。

6.2 所学技能

在本章，你应该学到了以下知识。

- ◆ 建立更高级的 TensorFlow 库，包括那些需要 Bazel 驱动的编译器。
- ◆ 处理文本数据。
- ◆ 将 RNNs 和 CNNs 应用到文本上，而非图像上。
- ◆ 利用保存的模型评估文本。
- ◆ 使用预构建的库来提取句子结构细节。
- ◆ 根据积极和消极的情绪将文本分成不同的类别。

6.3 总结

在本章，我们将神经网络知识应用于文本来理解语言。这是一个相当大的成就，因为如果它能完全自动化就可以扩大应用规模。即使特定的评估结果不正确，但从统计学角度来看，我们仍拥有一个使用相同构建模块构建的强大工具。

第 7 章
利用机器学习赚钱

到目前为止，我们主要使用 TensorFlow 进行了图像处理，并在一定程度上进行了文本序列处理。在本章中，我们将处理一种特定类型的表格数据：时间序列数据。

时间序列数据来自很多领域，通常这些领域都有一个共同点——唯一不断变化的字段是时间或序列字段。时间序列很常见，尤其是在经济、金融、保健、医学、环境工程和控制工程领域。我们将在本章深入探讨一些示例，需要记住的关键点是顺序。与之前章节不同，如果时间序列数据被打乱，那么将会损失其原有的含义。另外，还需要注意数据的可用性。如果我们只有到目前为止的数据，而无法获取更多的历史数据，那么也将无法产生更多的数据集——受限于基于时间的可用性。

在本章，我们将进入一个拥有大量数据的领域：金融世界。我们将探索对冲基金以及经验丰富的投资者可能会利用时间序列数据所做的一些事情。

在本章中，我们将学习以下主题。

- ◆ 时间序列数据及其特性。
- ◆ 投资公司在机器学习驱动的大量投资工作中使用的输入和方法。
- ◆ 金融时间序列数据及其获取方式，以及一些实时金融数据的获取。
- ◆ 改进的卷积神经网络在金融上的应用。

7.1 输入和方法

投资公司的内部自营交易组使用多种手段进行投资、交易和获利。对冲基金则使用了一种更广泛、更有趣、更复杂的投资方式。一些投资受大量的理性或非理性的思维影响，而其他的投资则大多通过量化被过滤器、算法或信号所驱动。这两种方式都很好，在本章中我们将关注后者。

这些量化的方法用到了很多技术。其中的一些技术如下所示。

- 基于评估的技术。
- 基于异常和信号的技术。
- 基于外部信号的技术。
- 基于过滤和分段的组群分析（cohort analysis）。

其中一些方法使用传统的机器学习技术，例如 K 最近邻法（K-Nearest Neighbors，KNN）、朴素贝叶斯法（Naive Bayes）和支持向量机（Support Vector Machines，SVM）。尤其是组群分析，它几乎是 KNN 方法的一种完美方案。

另一种常用的技术是情感分析（sentiment analysis）和基于人群情感的信号（crowd-sentiment-based signal）。在上一章中我们利用该技术评估了文本情感，并将文本段落分成几种基本类别：积极的、消极的、快乐的、生气的等。如果我们获取了更多的数据，并过滤出除涉及特定股票之外的所有内容，那么就可以得到股票的价格。因此现在我们需要拥有一个分布广泛（可能是全球性的）、容量大和速度快的文本来源。Twitter 通过 API 开放的数据，以及 Facebook 和其他社交媒体平台开放的数据都可以作为我们所需的文本来源。事实上，一些对冲基金使用了 Twitter 和 Facebook 所开放的整个数据，并试图从中获取公众对股市、市场部门、商品期货等方面的情绪。这是一种由 NLP 驱动的基于信号的外部投资策略，从业者使用这种策略来预测时间序列的方向和（或）强度。

在本章，我们将利用内部措施来预测时间序列上未来的入口，而这个内部措施本身也运用了时间序列。预测未来真实的入口实际上是一项非常困难的任务，而事实证明，我们完全没必要去预测。所以通常预测一个方向上的投资观点就足够了，如果这个投资观点结合了投资活动强度则更好。

对于很多类型的投资来说，即使是投资观点也不能提供百分之百的保证，其实只要在平均水平上能够确定正确或错误的概率哪个更大就足够了。这如同抛硬币判断正反的概率——如果长时间情况下你猜对的概率为 51%，并且你能够成千上万次地重复该游戏（这样足够从中获利），那么你赚取的将会更多。

这对基于机器学习的工作来说是一个好兆头，虽然我们可能无法对自己的回答怀有 100% 的信心，但能在统计学上拥有很好的预测能力。总之我们想要赚钱，因为即使是轻微的增长，经过成千上万次的循环也可能产生可观的收益。

获取数据

首先，我们来获取一些数据。在本章，我们将使用来自 Quandl 的数据。Quandl 通过一系列机制来提供很多股票数据。其中的一种简单的机制是通过 URL API 来实现的。

通过 Python 实现自动化相当容易。我们将使用以下代码来完成。

```
import requests

API_KEY = '<your_api_key>'

start_date = '2010-01-01'
end_date = '2015-01-01'
order = 'asc'
column_index = 4

stock_exchange = 'WIKI'
index = 'GOOG'

data_specs = 'start_date={}&end_date={}&order={}&column_index={}&api_key={}'
    .format(start_date, end_date, order, column_index, API_KEY)
base_url = "https://www.quandl.com/api/v3/datasets/{}/{}/data.json?" + data_specs
stock_data = requests.get(base_url.format(stock_exchange, index)).json()
```

因此，你可以把来自 WIKI/GOOG 的股票数据变量保存到 `stock_data` 变量中，即从格式化的 URL 中下载的 2010/01/01 到 2015/01/01 的数据。另外，变量 `column_index = 4` 用于告知服务器只获取历史上的收盘价。

 请注意，你可以在异步社区上找到本章的代码。

那么，这些收盘价到底是什么呢？我们知道，股票价格每天都在波动。它们会以某个特定的价格开盘，并在一天之内达到最高价和最低价，在当天结束的时候，它们会以某个特定的价格收盘。图 7-1 展示了在 1999-11-18～1999-12-17 某股票每天的价格是如何变化的（价格单位为美元）。

日期	开盘价	最高价	最低价	收盘价
1999-11-18	45.5	50	40	44
1999-11-19	42.94	43	39.81	40.38
1999-11-22	41.31	44	40.06	44
1999-11-23	42.5	43.63	40.25	40.25
1999-11-24	40.13	41.94	40	41.06
1999-11-26	40.88	41.5	40.75	41.19
1999-11-29	41	42.44	40.56	42.13
1999-11-30	42	42.94	40.94	42.19
1999-12-01	42.19	43.44	41.88	42.94
1999-12-02	43.75	45	43.19	44.13
1999-12-03	44.94	45.69	44.31	44.5
1999-12-06	45.25	46.44	45.19	45.75
1999-12-07	45.75	46	44.31	45.25
1999-12-08	45.25	45.63	44.81	45.19
1999-12-09	45.25	45.94	45.25	45.81
1999-12-10	45.69	45.94	44.75	44.75
1999-12-13	45.5	46.25	44.38	45.5
1999-12-14	45.38	45.38	42.06	43
1999-12-15	42	42.31	41	41.69
1999-12-16	42	48	42	47.25
1999-12-17	46.38	47.12	45.44	45.94

图 7-1

在股票开盘后，你可以进行投资并购买股票。在一天结束的时候，盈利还是亏损取决于你所购买的股票的收盘价。投资者使用不同的技术来预测在具体某天哪些股票有可能上涨，然后根据他们的分析来投资股票。

7.2　处理问题

在本章中，我们将根据其他时区（确保它们的收盘时间早于我们将要投资的股市）股市的涨跌情况，来确定股票价格是上涨还是下跌。我们将分析欧洲市场的数据，它们的收盘时间比美国股市大约早 3～4 小时。在 Quandl 中，我们将从以下欧洲市场获取数据。

- WSE/OPONEO_PL。
- WSE/VINDEXUS。
- WSE/WAWEL。
- WSE/WIELTON。

我们将预测美国市场 WIKI/SNPS 收盘价格的涨跌情况。

下载所有的市场数据，查看下载的数据以了解市场收盘价，然后修改数据，使其能够用于训练我们的网络。然后，查看该网络在我们假设的基础上的表现。

本章所使用的代码和分析技术受谷歌 Cloud Datalab Notebook 的启发。

操作步骤如下所示。

1. 下载所需数据并修改它。
2. 查看原始数据和修改后的数据。
3. 从修改后的数据中提取特征。
4. 准备训练并测试网络。
5. 构建网络。
6. 训练。
7. 测试。

7.2.1 下载和修改数据

在本节中，我们将从 codes 变量中的数据源下载数据，并将数据保存到 closings 数据帧中。此外，还将存储原始数据、scaled 数据和返回日志（log-return）。

```
codes = ["WSE/OPONEO_PL", "WSE/VINDEXUS", "WSE/WAWEL",
"WSE/WIELTON", "WIKI/SNPS"]
closings = pd.DataFrame()
for code in codes:
    code_splits = code.split("/")
    stock_exchange = code_splits[0]
    index = code_splits[1]
    stock_data = requests.get(base_url.format(stock_exchange,
    index)).json()
```

```
            dataset_data = stock_data['dataset_data']
            data = np.array(dataset_data['data'])
            closings[index] = pd.Series(data[:, 1].astype(float))
            closings[index + "_scaled"] = closings[index] /
             max(closings[index])
            closings[index + "_log_return"] = np.log(closings[index] /
closings[index].shift())
        closings = closings.fillna(method='ffill')   # Fill the gaps in data
```

我们对数据进行了缩放，以使股票值保持在 0～1，这有助于与其他股票值进行比较。它还有助于了解股票相对于其他市场的走势，并使分析更直观。

返回日志有助于得到相对于前一天的股价上涨和下跌的波动图。

7.2.2 查看数据

下面的代码片段将绘制我们下载和处理过的数据。

```
def show_plot(key="", show=True):
    fig = plt.figure()
    fig.set_figwidth(20)
    fig.set_figheight(15)
    for code in codes:
        index = code.split("/")[1]
        if key and len(key) > 0:
            label = "{}_{}".format(index, key)
        else:
            label = index
        _ = plt.plot(closings[label], label=label)

    _ = plt.legend(loc='upper right')
    if show:
        plt.show()

show = True
show_plot("", show=show)
show_plot("scaled", show=show)
show_plot("log_return", show=show)
```

图 7-2 展示了收盘价的原始市场数据。从图 7-2 可以看到，WAWEL 的值比其他市场的要大得多。

图 7-2

WAWEL 的收盘价在视觉上减弱了其他市场收盘价的变化趋势。因此，我们将对这些数据进行缩放，以便更好地查看，如图 7-3 所示。

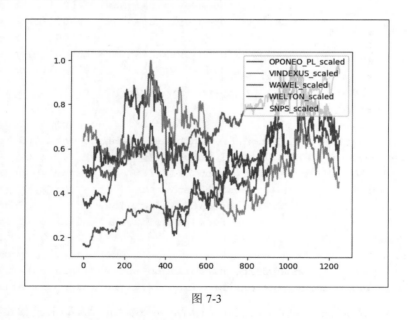

图 7-3

缩放后的市场值有助于更好地查看变化趋势。`log_return` 如图 7-4 所示。

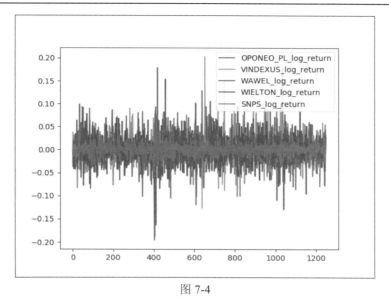

图 7-4

`log_return` 返回了市场的收盘价。

7.2.3 提取特征

现在,提取所需的特征来训练并测试我们的数据。

```
feature_columns = ['SNPS_log_return_positive',
'SNPS_log_return_negative']
for i in range(len(codes)):
    index = codes[i].split("/")[1]
    feature_columns.extend([
        '{}_log_return_1'.format(index),
        '{}_log_return_2'.format(index),
        '{}_log_return_3'.format(index)
    ])
features_and_labels = pd.DataFrame(columns=feature_columns)
closings['SNPS_log_return_positive'] = 0
closings.ix[closings['SNPS_log_return'] >= 0,
'SNPS_log_return_positive'] = 1
closings['SNPS_log_return_negative'] = 0
closings.ix[closings['SNPS_log_return'] < 0,
'SNPS_log_return_negative'] = 1
for i in range(7, len(closings)):
    feed_dict = {'SNPS_log_return_positive':
closings['SNPS_log_return_positive'].ix[i],
```

```
            'SNPS_log_return_negative':
closings['SNPS_log_return_negative'].ix[i]}
        for j in range(len(codes)):
            index = codes[j].split("/")[1]
            k = 1 if j == len(codes) - 1 else 0
            feed_dict.update({'{}_log_return_1'.format(index):
closings['{}_log_return'.format(index)].ix[i - k],
                '{}_log_return_2'.format(index):
closings['{}_log_return'.format(index)].ix[i - 1 - k],
                '{}_log_return_3'.format(index):
closings['{}_log_return'.format(index)].ix[i - 2 - k]})
        features_and_labels = features_and_labels.append(feed_dict,
ignore_index=True)
```

我们将所有的特征值和标签都存储在 `features_and_label` 变量中。`SNPS_log_return_positive` 和 `SNPS_log_return_negative` 键分别存储了日志返回的 SNPS 为正和为负时所对应的点。如果为正则返回 1，如果为负则返回 0。这两个键将充当网络的标签。

其他键则存储了其他市场过去 3 天的值（以及 SNPS 前 3 天的值，因为我们无法获取该市场当天的值）。

7.2.4　准备训练和测试

现在，将特征值划分为 train 子集和 test 子集。我们不会任意更改这些数据的排序，因为在金融市场的时间序列中，每天的数据都是按一定的规律产生的，我们必须遵循这种规律。即使对未来的数据进行训练，也无法通过过去的行为来验证训练是否有效，因为这样做毫无意义。我们总是对股票市场未来的表现感兴趣。

```
features = features_and_labels[features_and_labels.columns[2:]]
labels = features_and_labels[features_and_labels.columns[:2]]
train_size = int(len(features_and_labels) * train_test_split)
test_size = len(features_and_labels) - train_size
train_features = features[:train_size]
train_labels = labels[:train_size]
test_features = features[train_size:]
test_labels = labels[train_size:]
```

7.2.5　构建网络

训练时间序列的网络模型如下所示。

```
sess = tf.Session()
num_predictors = len(train_features.columns)
num_classes = len(train_labels.columns)
feature_data = tf.placeholder("float", [None, num_predictors])
actual_classes = tf.placeholder("float", [None, 2])
weights1 = tf.Variable(tf.truncated_normal([len(codes) * 3, 50],
stddev=0.0001))
biases1 = tf.Variable(tf.ones([50]))
weights2 = tf.Variable(tf.truncated_normal([50, 25],
stddev=0.0001))
biases2 = tf.Variable(tf.ones([25]))
weights3 = tf.Variable(tf.truncated_normal([25, 2], stddev=0.0001))
biases3 = tf.Variable(tf.ones([2]))
hidden_layer_1 = tf.nn.relu(tf.matmul(feature_data, weights1) +
biases1)
hidden_layer_2 = tf.nn.relu(tf.matmul(hidden_layer_1, weights2) +
biases2)
model = tf.nn.softmax(tf.matmul(hidden_layer_2, weights3) +
biases3)
cost = -tf.reduce_sum(actual_classes * tf.log(model))
train_op1 =
tf.train.AdamOptimizer(learning_rate=0.0001).minimize(cost)
init = tf.initialize_all_variables()
sess.run(init)
correct_prediction = tf.equal(tf.argmax(model, 1),
tf.argmax(actual_classes, 1))
accuracy = tf.reduce_mean(tf.cast(correct_prediction, "float"))
```

这只是一个包含两个隐藏层的简单网络。

7.2.6 训练

训练网络的代码如下所示。

```
for i in range(1, 30001):
    sess.run(train_op1, feed_dict={feature_data:
train_features.values,
            actual_classes:
train_labels.values.reshape(len(train_labels.values), 2)})
    if i % 5000 == 0:
        print(i, sess.run(accuracy, feed_dict={feature_data:
train_features.values,
                actual_classes:
train_labels.values.reshape(len(train_labels.values), 2)}))
```

7.2.7 测试

测试网络的代码如下所示。

```
feed_dict = {
    feature_data: test_features.values,
    actual_classes:
test_labels.values.reshape(len(test_labels.values), 2)
}
tf_confusion_metrics(model, actual_classes, sess, feed_dict)
```

7.3 更进一步

假设你刚刚训练了一个不错的分类器，并且它能够对市场进行一些预测，那么此时应该开始交易吗？与目前我们做过的其他机器学习项目类似，你需要在一个独立的测试集上进行测试。过去，我们通常会将数据分成以下 3 个数据集。

- 训练集。
- 开发集，也就是验证集。
- 测试集。

对于当前的工作我们也可以做类似的事情，不过金融市场提供了另一个资源——持续不断的数据流。

我们可以使用前面获取的数据源，并继续获取更多的数据。实际上，我们拥有一个不断扩展的、无形的数据集，这取决于我们使用数据的频率——如果操作的是每天的数据，那么实现该目标需要一点时间；如果是每小时或者每分钟的数据则更容易实现目标，因为我们能在较短时间内获取更多的数据。基于报价量的秒级数据通常更好。

此处使用现实货币可能会存在风险，所以大多数人通常都采用虚拟交易。本质上来说，系统几乎是实时运行的，且不会花费任何金钱，只需要记录系统实时运行时的操作即可。如果可行，那么下一步就可以使用真正的金钱进行实时交易（通常，只需使用少量的金钱测试该系统）。

7.4 个人的实际考虑

假设你训练了一个精确的分类器，且该分类器在盲数据集或实时数据集上表现出了良好的效果，那么现在你应该开始交易吗？虽然这是有可能的，但这并不容易。以下是部分原因。

- **历史分析与流数据**：这种历史数据通常被清理过且接近完美，但流数据并不具备这种优点。你需要通过编码来评估数据流，并别除潜在的不可靠数据。
- **买卖差价**：实际上市场中有两个价格，即买入的价格和卖出的价格。你不会按照市场价格同时买入和卖出股票（市场价格是买入价格和卖出价格的集合点，称为最终价格）。由于这个差价的存在，当你买入一支股票并立刻卖出时将会损失金钱，所以实际上，你已经在亏本了。
- **交易成本**：即使是每单交易小到 1 美元，交易成本仍然是一个我们需要克服的障碍。在我们的策略能够盈利之前，必须解决它。
- **税金**：这一点经常被人遗忘，可能是因为税金意味着净收益，这通常是一件好事。
- **退出能力**：在理论上你可以卖出股票，但并不意味着真实存在一个市场可以卖出你持有的股票，即使存在这样的市场，也很可能不适用于你所有的股票。这就需要更多的代码。这时，可以考虑出价、价格的交易量，以及股票的数量。
- **交易量和流动性**：信号告诉你要购买股票，但并不意味着市场上有足够的股票来供你购买。你看到的可能只是部分的股票交易量，实际的股票交易量仅仅露出了很小一部分。考虑到这一点，同样需要更多的代码。
- **与交易 APIs 的集成**：调用库很简单，但涉及金钱时则变得不再简单，此时需要交易协议、API 协议等。但是，成千上万的人已经做了这一点，并且对于那些寻求 APIs 来买入和卖出股票的人来说，盈透证券（Interactive Brokers）是最受欢迎的经纪业务之一。方便的是，他们也有一种 API 来提供市场数据。

7.5 所学技能

在本章中，你应该学到了以下技能。

- 理解时间序列数据。
- 为时间序列数据建立一个管道。
- 整合主要数据。
- 创建训练集和测试集。

- 综合考虑实际情况。

7.6 总结

针对金融数据的机器学习与我们使用的很多其他数据的机器学习并没有什么不同，事实上，我们使用的网络和用于其他数据集的网络是一样的。虽然还可以有其他选择，但是常规方法完全相同。特别是在交易金钱时，我们会发现实现代码会比实际的机器学习代码量要大得多。

在下一章中，我们将学习如何利用机器学习来实现医疗目标。

第 8 章
医疗应用

到目前为止，我们已经使用深度网络来处理图像、文本和时间序列了。虽然大多数示例比较有趣且富有意义，但它们未达到企业级水平。现在，我们将处理一个企业级问题：医疗诊断。之所以进行这个企业级问题的处理研究，是因为医疗数据通常具有其他大型企业未涉及的特性，即专有数据格式、天然的大规模、棘手的分类数据和不规则的特征。

本章将涉及以下主题。

◆ 医学影像文件及其独特性。

◆ 处理大型图像文件。

◆ 从典型医疗文件中提取分类数据。

◆ 应用非医疗数据"预训练"的网络。

◆ 扩展训练以适应医疗数据的规模。

获取医疗数据本身就是一项挑战，因此我们将利用一个所有相关从业人员都熟悉的网站——Kaggle。Kaggle 上面有很多可以免费获取的医疗数据集，但大多数需要进行相应的注册才能访问。另外，很多数据集只在医学图像处理领域的特定子社区发布，并且有特定的提交流程。Kaggle 也许是你能获取重要医疗图像数据集和非医疗图像数据集的最规范化的来源之一。在本章中，我们将特别关注 Kaggle 的糖尿病视网膜病变检测（Diabetic Retinopathy Detection）竞赛。在该竞赛中，参赛者利用 Kaggle 网站提供的大量眼底图，对其进行相应的图像处理，并利用机器学习算法对模型进行训练，最后利用测试集对训练得到的模型进行测试，根据测试结果判断该模型的优秀程度。

数据集可从 Kaggle 官网上下载。这个数据集包含一个训练集和一个盲测试集。训练集用于训练我们的网络，而测试集则用于测试网络。最后将我们的测试结果提交到 Kaggle 网站上。

由于数据量相当庞大（训练集 32 GB，测试集 49 GB），因此二者都被分成了多个大小约为 8GB 的 ZIP 文件。

此处的测试集是盲测试集，即我们不知道它们的标签。这是为了确保训练网络能够公平地提交测试集结果。

训练集的标签在 trainLabels.csv 文件中。

8.1 挑战

在深入介绍代码之前，我们需要了解大多数机器学习算法是如何实现分类或排名目标的。在很多情况下，分类本身就是一种排名，因为我们最终选择了排名等级最高的分类（通常是一个概率值）。我们涉及的医疗图像并没什么不同——我们会将这些图像分到以下两类之一。

- 疾病状态/正值。

- 正常状态/负值。

或者，我们会把它们分成多个种类或对它们进行排序。在糖尿病视网膜病变示例中，分类级别排序如下。

- 0 级：没有糖尿病视网膜病变。

- 1 级：轻微。

- 2 级：中度。

- 3 级：严重。

- 4 级：大面积的糖尿病视网膜病变。

分类级别的数值通常叫作得分。Kaggle 为参赛者提供超过 32 GB 的训练数据，其中包含 35 000 多张图片，而测试数据则更大——49 GB。该挑战的目标是使用已知得分在 35 000 多张图片上进行训练，并获取在测试集上的得分。训练标签如表 8-1 所示。

表 8-1

图　　像	分 类 级 别
10_left	0

续表

图像	分类级别
10_right	0
13_left	0
13_right	0
15_left	1
15_right	2
16_left	4
16_right	4
17_left	0
17_right	1

背景知识：糖尿病视网膜病变是一种视网膜疾病，存在于眼睛内部，所以我们有左眼和右眼的得分，可以将两者当作独立的训练数据，或者对单个病人进行整体考虑。接下来就开始实现吧。

到目前为止，你可能已经熟悉了获取一组数据并将其分割成训练数据块、验证数据块和测试数据块的步骤。对于之前使用的一些标准数据集来说，这种方式效果不错，但是当前的数据集是竞赛的一部分，并且是公开评审的，所以并不知道答案。这能很好地反映真实生活。一个好的方面是，大多数 Kaggle 竞赛会让你给出一个答案，并告诉你你的总分，这有助于模型进行学习并了解改进方向。此外，这也有助于参赛者和社区知晓哪些用户做得比较好。

由于测试集标签是未知的，因此我们需要对之前所做的两项工作进行如下调整。

◆ 需要有一个内部开发和迭代的过程（可能会将训练集分成训练集、验证集和测试集），还需要一个外部测试的过程（可能会确定一个工作良好的设置，然后在盲测试集上运行，或者在整个训练集上再次训练）。

◆ 需要以一种特定的格式做一个正式的提案，并将其提交给独立的评审者（在本例中为 Kaggle），并相应地评估进展。表 8-2 展示了一个提交的示例，它看起来非常类似于训练标签文件。

表 8-2

图像	分类级别
44342_left	0
44342_right	1

续表

图 像	分类级别
44344_left	2
44344_right	2
44345_left	0
44345_right	0
44346_left	4
44346_right	3
44350_left	1
44350_right	1
44351_left	4
44351_right	4

8.2 数据

接下来开始查看数据。打开示例文件，可以看到，这些既不是 28×28 的手写字符块，也不是 64×64 的猫脸图标，而是来自现实世界的真实数据集。事实上，甚至连图像的尺寸都不一致。

你会发现单边尺寸从 2 000 像素到 5 000 像素不等。这就将我们带到了第一个现实生活任务——创建一个训练管道。该管道通过一系列的步骤筛选数据，最终产生一组干净、一致的数据。

8.3 管道

我们将会理智地处理该任务。在 TensorFlow 库中，谷歌使用不同网络创建了很多管道模型结构。在这里，我们将使用其中的一种模型结构和网络，并根据实际需要修改代码。

这种方式比较好，因为无须浪费时间从头开始创建管道，也无须担忧如何整合 TensorBoard 可视化组件，因为谷歌管道模型中已经包含了它。

我们将使用这里的一个管道模型，可在 GitHub 中搜索 TensorFlow/models。

正如你所看到的，GitHub 中有很多在 TensorFlow 里创建的不同模型。你可以深入研究一些与自然语言处理（NLP）、递归神经网络及其他主题相关的模型。如果你想理解复杂模型，那么这是一个很好的起点。

在本章中，我们将使用 Tensorflow-Slim 图像分类模型库。

GitHub 网站上有很多关于如何使用该库的详细信息。它们还介绍了如何在分布式环境中使用该库，如何利用多个 GPU 来获取更快的训练速度，甚至包括如何部署生产环境。

使用这种方式的优势是，它们为你提供了预训练的模型快照，这可以极大地减少网络的训练时间。所以，为了达到合适的训练水平，即使你的 GPU 速度很慢，也无须花费数周时间来训练这么大规模的网络。

这就是模型的微调，在这里只需提供一个不同的数据集，并告诉网络重新初始化网络的最后一层，以便重新训练它们。此外，还需要告诉它数据集中有多少个输出标签类别。在我们的示例中，有 5 种不同的类别来标识不同级别的糖尿病视网膜病变（DR）。

正如在前面所看到的，它们提供了很多类型的预训练模型供我们使用，这些模型经过了 ImageNet 数据集的训练。ImageNet 是一个包含 1 000 个类别的标准数据集，大小接近 500 GB。

8.3.1 理解管道

首先，将模型仓库复制到计算机上。

git clone https://GitHub 网址/tensorflow/models/

现在，我们将深入了解从谷歌的模型仓库中获取的管道。

如果查看仓库中路径前缀 model/research/slim 下的文件夹，将会看到名为 datasets、deployment、nets、preprocessing 和 scripts 的文件夹，一些与生成模型相关的文件、训练和测试管道，与训练 ImageNet 数据集相关的文件，以及一个名为 flowers 的数据集。

此处将使用 download_and_convert_data.py 构建 DR 数据集。这个图像分类模型库基于 slim 库创建。在本章中，我们将微调 nets/inception_v3.py（在本章稍后部分将对网络规范及其概念进行更多讨论）中定义的初始网络，其中包含了计算损失函数、添加不同操作、构造网络等。最后，train_image_classifier.py 和 eval_image_classifier.py 文件包含了为网络创建的训练管道和测试管道的通用程序。

在本章中，由于网络的复杂性，所以我们会使用一个基于 GPU 的管道来训练网络。如果想知道如何在计算机上安装支持 GPU 的 TensorFlow，可以参考本书第 12 章的内容。另外，为了确保能够正常运行代码，计算机应该保留 120 GB 的空间。读者可以在本书代码文件的 Chapter_08 文件夹中找到最终的代码文件。

8.3.2 准备数据集

现在，开始为网络准备数据集。

对于该初始网络，我们将使用 TFRecord 类来管理数据集。预处理之后的输出数据集文件是归档文件（protofile），它可以通过 TFRecord 读取，归档文件将数据以序列化的格式进行存储，以此提高读取速度。每个归档文件中都存储了一些信息，如图像大小和格式等信息。

这样做的原因是：数据集的规模太大，它将占用大量的空间，从而无法加载到内存（RAM）中。因此，为了有效使用 RAM，必须分批加载图像，并将之前加载且目前未使用的图像删除。

该网络接收的输入大小为 299×299。所以，我们需要寻找一种方法先将图像大小缩减为 299×299，以得到一个具有一致性图像的数据集。

在缩小图像之后，我们将创建一些归档文件，之后可以将这些文件输入到训练网络中。

首先，需要从 Kaggle 官网下载糖尿病视网膜病变检测的 5 个训练集文件和标签文件。

但是，Kaggle 只允许你通过一个账号下载训练集 ZIP 文件，所以下载数据集文件（类似前几章）的过程将无法实现自动化。

现在，假设已经下载了所有的 5 个训练集文件和标签文件，并将它们存储在一个名为 diabetic 的文件夹中。此时，diabetic 文件夹的结构如下所示。

- train.zip.001。
- train.zip.002。
- train.zip.003。
- train.zip.004。
- train.zip.005。

- trainLabels.csv.zip。

为了简化项目,使用压缩软件手动提取。提取完成后,diabetic 文件夹的结构如下所示。

- train。
- 10_left.jpeg。
- 10_right.jpeg。
- …
- trainLabels.csv。
- train.zip.001。
- train.zip.002。
- train.zip.003。
- train.zip.004。
- train.zip.005。
- trainLabels.csv.zip。

在本例中,train 文件夹包含.zip 文件中所有的图像,而 trainLabels.csv 包含每张图像的真实值标签。

模型仓库的作者提供了一些示例代码来处理一些典型的图像分类数据集。糖尿病问题可以使用同样的方法来解决。因此,我们可以利用处理其他数据集(例如 flowers 数据集或 MNIST 数据集)的代码。本书的代码库提供了修改过的能够处理糖尿病问题的代码。

你需要复制代码库,并导航到 chapter_08 文件夹。可以按如下方式运行 download_and_convert_data.py 文件。

```
python download_and_convert_data.py --dataset_name diabetic --dataset_dir D:\\datasets\\diabetic
```

在本例中,将 `dataset_name` 设置为 `diabetic`,而 `dataset_dir` 是包含 trainLabels.csv 和 train 文件夹的文件夹。

代码运行应该没有任何问题,先将数据集预处理成一种合适的格式(299×299),并在新建的名为 tfrecords 的文件夹中创建一些 TFRecord 文件。图 8-1 展示了 tfrecords 文件夹中的内容。

图 8-1

8.3.3 解释数据准备

现在，我们来到了数据预处理的编码部分。从现在开始，我们将展示在最初的 tensorflow/models 仓库中所做的修改。基本上，需要从处理 flowers 数据集的代码开始进行修改，以满足我们的需要。

在 download_and_convert_data.py 文件中，我们在开头处添加了一行新代码。

```
from datasets import download_and_convert_diabetic
and a new else-if clause to process the dataset_name "diabetic" at line 69:
    elif FLAGS.dataset_name == 'diabetic':
        download_and_convert_diabetic.run(FLAGS.dataset_dir)
```

利用此代码，就可以调用 datasets 文件夹内 download_and_convert_diabetic.py 中的 run 方法。这是分离多个数据集以预处理代码的一种非常简单的方法，而对于图像分类库中的其他部分，我们可以无须修改而加以利用。

download_and_convert_diabetic.py 文件是 download_and_convert_flowers.py 的一个副本，我们对其做了一些修改以用于糖尿病数据集。

在 download_and_convert_diabetic.py 文件中的 run 方法中，我们做了以下修改。

```
def run(dataset_dir):
  """Runs the download and conversion operation.

  Args:
    dataset_dir: The dataset directory where the dataset is stored.
  """
  if not tf.gfile.Exists(dataset_dir):
      tf.gfile.MakeDirs(dataset_dir)

  if _dataset_exists(dataset_dir):
      print('Dataset files already exist. Exiting without re-creating them.')
      return

  # Pre-processing the images.
  data_utils.prepare_dr_dataset(dataset_dir)
  training_filenames, validation_filenames, class_names = _get_filenames_and_classes(dataset_dir)
  class_names_to_ids = dict(zip(class_names, range(len(class_names))))

  # Convert the training and validation sets.
  _convert_dataset('train', training_filenames, class_names_to_ids, dataset_dir)
  _convert_dataset('validation', validation_filenames, class_names_to_ids, dataset_dir)

  # Finally, write the labels file:
  labels_to_class_names = dict(zip(range(len(class_names)), class_names))
  dataset_utils.write_label_file(labels_to_class_names, dataset_dir)

  print('\nFinished converting the Diabetic dataset!')
```

在该代码中，我们使用了 data_utils 包的 prepare_dr_dataset，它存在于本书仓库的根目录中。稍后将讨论该方法。修改 _get_filenames_and_classes 方法来返回训练和验证文件名。最后几行代码与 flowers 数据集示例相同。

```python
def _get_filenames_and_classes(dataset_dir):
    train_root = os.path.join(dataset_dir, 'processed_images', 'train')
    validation_root = os.path.join(dataset_dir, 'processed_images',
    'validation')
    class_names = []
    for filename in os.listdir(train_root):
        path = os.path.join(train_root, filename)
        if os.path.isdir(path):
            class_names.append(filename)

    train_filenames = []
    directories = [os.path.join(train_root, name) for name in
    class_names]
    for directory in directories:
        for filename in os.listdir(directory):
            path = os.path.join(directory, filename)
            train_filenames.append(path)

    validation_filenames = []
    directories = [os.path.join(validation_root, name) for name in
    class_names]
    for directory in directories:
        for filename in os.listdir(directory):
            path = os.path.join(directory, filename)
            validation_filenames.append(path)
    return train_filenames, validation_filenames, sorted(class_names)
```

在前面的方法中，我们在 processed_images/train 和 processed/validation 文件夹中找到了所有的文件名，里面包含了在 data_utils.prepare_dr_dataset 函数中预处理的图像。

我们在 **data_utils.py** 文件中编写了 prepare_dr_dataset(dataset_dir) 函数，它负责数据的预处理。

首先，定义必要的变量以链接到数据。

```python
num_of_processing_threads = 16
dr_dataset_base_path = os.path.realpath(dataset_dir)
unique_labels_file_path = os.path.join(dr_dataset_base_path,
"unique_labels_file.txt")
processed_images_folder = os.path.join(dr_dataset_base_path,
"processed_images")
num_of_processed_images = 35126
train_processed_images_folder =
```

```
os.path.join(processed_images_folder, "train")
validation_processed_images_folder =
os.path.join(processed_images_folder, "validation")
num_of_training_images = 30000
raw_images_folder = os.path.join(dr_dataset_base_path, "train")
train_labels_csv_path = os.path.join(dr_dataset_base_path,
"trainLabels.csv")
```

变量 num_of_processing_threads 用于指定预处理数据集时使用的线程数量。我们将使用一个多线程环境来更快地预处理数据。在预处理过程中，我们指定了一些目录路径，以使不同的文件夹包含我们的数据。

我们将提取图像的原始形式，并对其进行预处理，使其规格和大小合适且一致，再利用文件 download_and_convert_diabetic.py 中的 _convert_dataset 函数从处理过的图像中生成 tfrecords 文件。然后，将这些 tfrecords 文件输入训练网络和测试网络。

正如 8.3.2 节所述，我们已经提取了数据集文件和标签文件。现在，所有的数据已经被提取并保存在我们的计算机里，那么接下来就可以处理图像了。DR 数据集中一个典型的图像如图 8-2 所示。

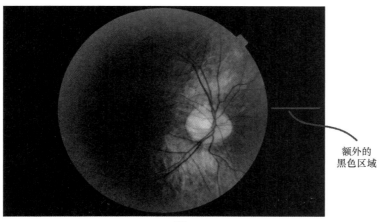

图 8-2

处理图像要做的就是移除这个额外的黑色区域，因为它对我们的网络来说不是必要的。这一步将减少图像中不必要的信息。在此之后，此图像将被缩放为一个 299×299 的 JPG 图像文件。

我们将对所有训练数据集重复该过程。

裁剪黑色图像边界的函数如下所示。

```python
def crop_black_borders(image, threshold=0):
    """Crops any edges below or equal to threshold

    Crops blank image to 1x1.

    Returns cropped image.

    """
    if len(image.shape) == 3:
        flatImage = np.max(image, 2)
    else:
        flatImage = image
    assert len(flatImage.shape) == 2

    rows = np.where(np.max(flatImage, 0) > threshold)[0]
    if rows.size:
        cols = np.where(np.max(flatImage, 1) > threshold)[0]
        image = image[cols[0]: cols[-1] + 1, rows[0]: rows[-1] + 1]
    else:
        image = image[:1, :1]

    return image
```

该函数获取图像和灰度阈值，当图像像素值低于该阈值时，该函数将会删除图像周围的黑色边框。

因为程序是在多线程环境中运行的，所以我们将批量处理这些图像。批量处理图像使用下面的代码。

```python
def process_images_batch(thread_index, files, labels, subset):

    num_of_files = len(files)

    for index, file_and_label in enumerate(zip(files, labels)):
        file = file_and_label[0] + '.jpeg'
        label = file_and_label[1]

        input_file = os.path.join(raw_images_folder, file)
        output_file = os.path.join(processed_images_folder, subset,
            str(label), file)
```

```
        image = ndimage.imread(input_file)
        cropped_image = crop_black_borders(image, 10)
        resized_cropped_image = imresize(cropped_image, (299, 299, 3),
        interp="bicubic")
        imsave(output_file, resized_cropped_image)

        if index % 10 == 0:
            print("(Thread {}): Files processed {} out of
            {}".format(thread_index, index, num_of_files))
```

thread_index 表示该函数被调用时的线程 ID。下述函数定义了批量处理图像的多线程环境。

```
def process_images(files, labels, subset):
    # Break all images into batches with a [ranges[i][0], ranges[i]
    [1]].
    spacing = np.linspace(0, len(files), num_of_processing_threads +
    1).astype(np.int)
    ranges = []
    for i in xrange(len(spacing) - 1):
        ranges.append([spacing[i], spacing[i + 1]])

    # Create a mechanism for monitoring when all threads are finished.
    coord = tf.train.Coordinator()

    threads = []
    for thread_index in xrange(len(ranges)):
        args = (thread_index, files[ranges[thread_index]
        [0]:ranges[thread_index][1]],
                labels[ranges[thread_index][0]:ranges[thread_index]
                [1]],
                subset)
        t = threading.Thread(target=process_images_batch, args=args)
        t.start()
        threads.append(t)

    # Wait for all the threads to terminate.
    coord.join(threads)
```

为了得到所有线程的最终结果，我们使用了一个 TensorFlow 类 tf.train.Coordinator()，它的 join 函数负责处理所有线程的最后接入点。

针对线程，我们使用 `threading.Thread(target,args)`，其中参数 `target` 指定调用的函数，参数 `args` 指定目标函数的参数。

现在，处理训练图像。将训练数据集分成一个训练集（30 000 张图片）和一个验证集（5 126 张图片）。

整个预处理过程如下。

```
def process_training_and_validation_images():
    train_files = []
    train_labels = []

    validation_files = []
    validation_labels = []

    with open(train_labels_csv_path) as csvfile:
        reader = csv.DictReader(csvfile)
        for index, row in enumerate(reader):
            if index < num_of_training_images:
                train_files.extend([row['image'].strip()])
                train_labels.extend([int(row['level'].strip())])
            else:
                validation_files.extend([row['image'].strip()])
                validation_labels.extend([int(row['level'].strip())])

    if not os.path.isdir(processed_images_folder):
        os.mkdir(processed_images_folder)

    if not os.path.isdir(train_processed_images_folder):
        os.mkdir(train_processed_images_folder)

    if not os.path.isdir(validation_processed_images_folder):
        os.mkdir(validation_processed_images_folder)

    for directory_index in range(5):
        train_directory_path =
 os.path.join(train_processed_images_folder,
 str(directory_index))
        valid_directory_path =
 os.path.join(validation_processed_images_folder,
 str(directory_index))
```

```
            if not os.path.isdir(train_directory_path):
                os.mkdir(train_directory_path)

            if not os.path.isdir(valid_directory_path):
                os.mkdir(valid_directory_path)

    print("Processing training files...")
    process_images(train_files, train_labels, "train")
    print("Done!")

    print("Processing validation files...")
    process_images(validation_files, validation_labels,
    "validation")
    print("Done!")

    print("Making unique labels file...")
    with open(unique_labels_file_path, 'w') as unique_labels_file:
        unique_labels = ""
        for index in range(5):
            unique_labels += "{}\n".format(index)
        unique_labels_file.write(unique_labels)

    status = check_folder_status(processed_images_folder,
    num_of_processed_images,
    "All processed images are present in place",
    "Couldn't complete the image processing of training and
    validation files.")

    return status
```

现在，分析用于整理数据的最后一个方法，即 download_and_convert_diabetic.py 文件中的 _convert_dataset 方法。

```
    def _get_dataset_filename(dataset_dir, split_name, shard_id):
        output_filename = 'diabetic_%s_%05d-of-%05d.tfrecord' % (
            split_name, shard_id, _NUM_SHARDS)
        return os.path.join(dataset_dir, output_filename)
    def _convert_dataset(split_name, filenames, class_names_to_ids,
    dataset_dir):
        """Converts the given filenames to a TFRecord dataset.

        Args:
```

```
        split_name: The name of the dataset, either 'train' or
        'validation'.
        filenames: A list of absolute paths to png or jpg images.
        class_names_to_ids: A dictionary from class names (strings)
to
        ids
          (integers).
        dataset_dir: The directory where the converted datasets are
        stored.
    """
    assert split_name in ['train', 'validation']
    num_per_shard = int(math.ceil(len(filenames) /
    float(_NUM_SHARDS)))

    with tf.Graph().as_default():
        image_reader = ImageReader()

        with tf.Session('') as sess:

            for shard_id in range(_NUM_SHARDS):
                output_filename = _get_dataset_filename(
                    dataset_dir, split_name, shard_id)

                with tf.python_io.TFRecordWriter(output_filename)
                as
                tfrecord_writer:
                    start_ndx = shard_id * num_per_shard
                    end_ndx = min((shard_id + 1) * num_per_shard,
                    len(filenames))
                    for i in range(start_ndx, end_ndx):
                        sys.stdout.write('\r>> Converting image
                        %d/%d shard %d' % (
                            i + 1, len(filenames), shard_id))
                        sys.stdout.flush()

                        # Read the filename:
                        image_data =
                        tf.gfile.FastGFile(filenames[i], 'rb').read()
                        height, width =
                        image_reader.read_image_dims(sess, image_data)

                        class_name =
```

```
                os.path.basename(os.path.dirname(filenames[i]))
                        class_id = class_names_to_ids[class_name]

                        example = dataset_utils.image_to_tfexample(
                            image_data, b'jpg', height, width,
                            class_id)
                    tfrecord_writer.write(example.SerializeToString())

                    sys.stdout.write('\n')
                    sys.stdout.flush()
```

在前面的函数中，我们获取了图像文件名，然后将它们存储在 tfrecord 文件中。我们还将训练文件和验证文件分割成多个 tfrecord 文件，而不是每个分割集只使用一个文件。

现在，随着数据处理的结束，我们将数据集形式化为一个 slim.dataset 实例，它是 Tensorflow Slim 库中的一个数据集类。在 datasets/diabetic.py 文件中，你将会看到一个名为 get_split 的方法，如下所示。

```
_FILE_PATTERN = 'diabetic_%s_*.tfrecord'
SPLITS_TO_SIZES = {'train': 30000, 'validation': 5126}
_NUM_CLASSES = 5
_ITEMS_TO_DESCRIPTIONS = {
    'image': 'A color image of varying size.',
    'label': 'A single integer between 0 and 4',
}
def get_split(split_name, dataset_dir, file_pattern=None,
reader=None):
    """Gets a dataset tuple with instructions for reading flowers.
    Args:
      split_name: A train/validation split name.
      dataset_dir: The base directory of the dataset sources.
      file_pattern: The file pattern to use when matching the dataset
sources.
          It is assumed that the pattern contains a '%s' string so that
the split
         name can be inserted.
      reader: The TensorFlow reader type.
    Returns:
      A 'Dataset' namedtuple.
    Raises:
      ValueError: if 'split_name' is not a valid train/validation
split.
```

```python
    """
    if split_name not in SPLITS_TO_SIZES:
        raise ValueError('split name %s was not recognized.' % split_name)

    if not file_pattern:
        file_pattern = _FILE_PATTERN
    file_pattern = os.path.join(dataset_dir, file_pattern % split_name)

    # Allowing None in the signature so that dataset_factory can use the default.
    if reader is None:
        reader = tf.TFRecordReader

    keys_to_features = {
        'image/encoded': tf.FixedLenFeature((), tf.string, default_value=''),
        'image/format': tf.FixedLenFeature((), tf.string, default_value='png'),
        'image/class/label': tf.FixedLenFeature(
            [], tf.int64, default_value=tf.zeros([], dtype=tf.int64)),
    }
    items_to_handlers = {
        'image': slim.tfexample_decoder.Image(),
        'label': slim.tfexample_decoder.Tensor('image/class/label'),
    }
    decoder = slim.tfexample_decoder.TFExampleDecoder(
        keys_to_features, items_to_handlers)

    labels_to_names = None
    if dataset_utils.has_labels(dataset_dir):
      labels_to_names = dataset_utils.read_label_file(dataset_dir)

    return slim.dataset.Dataset(
        data_sources=file_pattern,
        reader=reader,
        decoder=decoder,
        num_samples=SPLITS_TO_SIZES[split_name],
        items_to_descriptions=_ITEMS_TO_DESCRIPTIONS,
        num_classes=_NUM_CLASSES,
        labels_to_names=labels_to_names)
```

前面的方法将会在训练程序和评估程序中被调用。我们将创建一个带有 tfrecord 文件信息的 slim.dataset 实例，这样它就可以自动执行二进制文件的解析工作了。此外，利用 Tensorflow Slim 对 DatasetDataProvider 的支持，我们也可以使用 slim.dataset.Dataset 来并行读取数据集，从而可以增加训练程序和评估程序。

在开始训练之前，需要从 Tensorflow Slim 图像分类库中下载预训练的模型 Inception V3，这样就无须从头开始训练，直接利用 Inception V3 的性能即可。

预训练的模型可从 https:/ /GitHub 网址/ tensorflow/ models/ tree/ master/ research/ slim#Pretrained 下载。

在本章中，我们将使用 Inception V3，所以需要下载 inception_v3_2016_08_28.tar.gz 文件，并从中提取出检查点文件 inception_v3.ckpt。

8.3.4 训练流程

现在，开始训练和评估我们的模型。

训练脚本在 train_image_classifer.py 文件中。因为我们一直跟随该库的工作流程，所以可以原封不动地使用该文件，并使用下面的指令运行训练流程。

```
python train_image_classifier.py --
train_dir=D:\datasets\diabetic\checkpoints --dataset_name=diabetic --
dataset_split_name=train --dataset_dir=D:\datasets\diabetic\tfrecords --
model_name=inception_v3 --
checkpoint_path=D:\datasets\diabetic\checkpoints\inception_v3\inception_v3.
ckpt --checkpoint_exclude_scopes=InceptionV3/Logits,InceptionV3/AuxLogits -
-trainable_scopes=InceptionV3/Logits,InceptionV3/AuxLogits --
learning_rate=0.000001 --learning_rate_decay_type=exponential
```

在上述设置中，我们需要通宵运行训练流程。现在，通过验证流程来运行训练的模型，以查看其效果。

8.3.5 验证流程

可以使用以下命令来运行验证流程。

```
python eval_image_classifier.py --alsologtostderr --
checkpoint_path=D:\datasets\diabetic\checkpoints\model.ckpt-92462 --
dataset_name=diabetic --dataset_split_name=validation --
dataset_dir=D:\datasets\diabetic\tfrecords --model_name=inception_v3
```

如图 8-3 所示，目前的准确率约为 75%。在 8.4 节中，我们将提供一些提高准确率的思路。

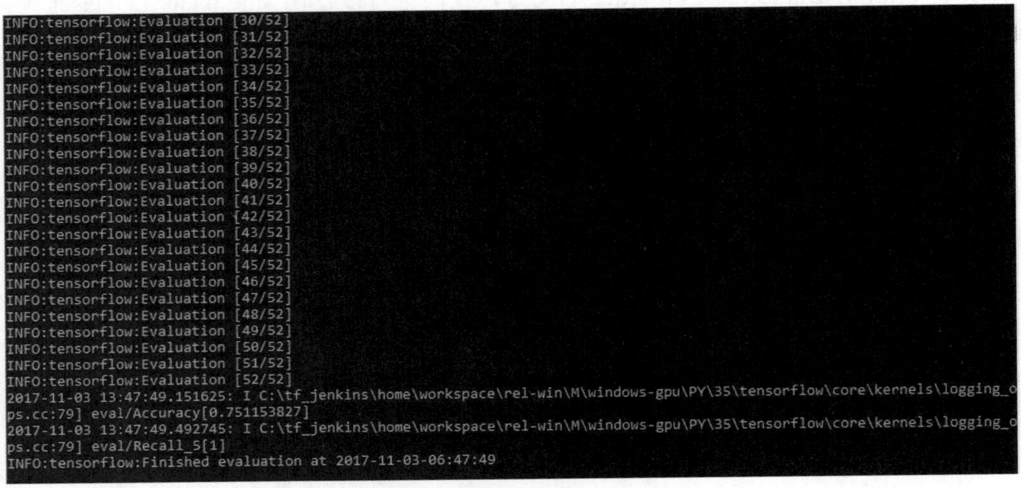

```
INFO:tensorflow:Evaluation [30/52]
INFO:tensorflow:Evaluation [31/52]
INFO:tensorflow:Evaluation [32/52]
INFO:tensorflow:Evaluation [33/52]
INFO:tensorflow:Evaluation [34/52]
INFO:tensorflow:Evaluation [35/52]
INFO:tensorflow:Evaluation [36/52]
INFO:tensorflow:Evaluation [37/52]
INFO:tensorflow:Evaluation [38/52]
INFO:tensorflow:Evaluation [39/52]
INFO:tensorflow:Evaluation [40/52]
INFO:tensorflow:Evaluation [41/52]
INFO:tensorflow:Evaluation [42/52]
INFO:tensorflow:Evaluation [43/52]
INFO:tensorflow:Evaluation [44/52]
INFO:tensorflow:Evaluation [45/52]
INFO:tensorflow:Evaluation [46/52]
INFO:tensorflow:Evaluation [47/52]
INFO:tensorflow:Evaluation [48/52]
INFO:tensorflow:Evaluation [49/52]
INFO:tensorflow:Evaluation [50/52]
INFO:tensorflow:Evaluation [51/52]
INFO:tensorflow:Evaluation [52/52]
2017-11-03 13:47:49.151625: I C:\tf_jenkins\home\workspace\rel-win\M\windows-gpu\PY\35\tensorflow\core\kernels\logging_o
ps.cc:79] eval/Accuracy[0.751153827]
2017-11-03 13:47:49.492745: I C:\tf_jenkins\home\workspace\rel-win\M\windows-gpu\PY\35\tensorflow\core\kernels\logging_o
ps.cc:79] eval/Recall_5[1]
INFO:tensorflow:Finished evaluation at 2017-11-03-06:47:49
```

图 8-3

现在，利用 TensorBoard 来可视化训练过程。

8.3.6 利用 TensorBoard 可视化训练过程

本节将使用 TensorBoard 来可视化训练过程。

首先，需要将命令行目录切换为包含检查点的文件夹。在我们的示例中，该文件夹是前面命令中的 `train_dir` 参数值，即 D:\datasets\diabetic\checkpoints。然后，运行以下命令。

```
tensorboard -logdir .
```

当网络运行 TensorBoard 时，输出如图 8-4 所示。

图 8-4 显示了包含训练网络 RMS 支持优化器的节点，以及该节点包含的用于处理 DR 分类输出的一些逻辑。

图 8-5 显示了输入的图像，以及对它们的预处理和修改情况。

8.3 管道　131

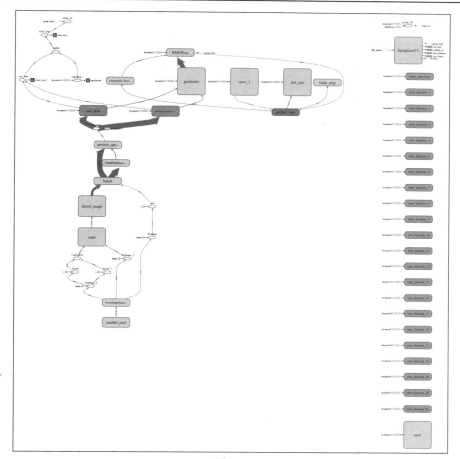

图 8-4

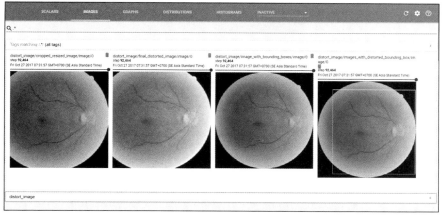

图 8-5

训练过程中的网络输出如图 8-6 所示。

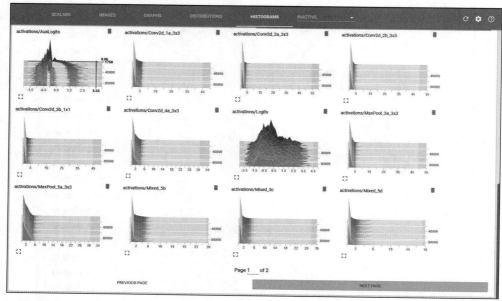

图 8-6

图 8-7 描述了训练过程中网络的总体原始损失。

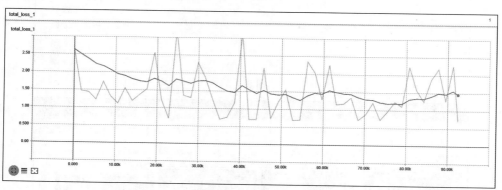

图 8-7

Inception 网络

Inception 网络背后的主要涵义是在单个网络层中将不同的卷积组合在一起。该组合是通过组合 7×7、5×5、3×3 和 1×1 的卷积并传递到下一层来实现的。这样就可以提取出更多的网络特征，从而获得更高的准确率。图 8-8 所示的谷歌 Inception V3 网络图展示了这一点。你可以在 chapter_08/nets/inception_v3.py 中访问其代码。

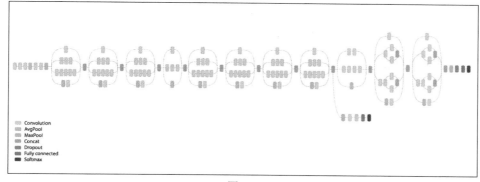

图 8-8

8.4 更进一步

Inception V3 网络在验证集上运行得到的准确率为 75%。因为该网络应用具有临界性，所以这个结果并不是很好。在医学上，一个人的身体状况时刻影响着自身安危，所以没有多少可以犯错的余地。

为了提高准确率，需要定义一个不同的评价标准。

此外，还需要平衡数据集。我们现在拥有的是一个不平衡的数据集，其中患该病的病人数量比普通病人的数量要少得多。因此，网络对普通病人的特征更加敏感，而对患该病的病人的特征的敏感性要低得多。

为了解决这个问题，我们可以对数据集进行 SMOTE 处理。SMOTE 代表合成少数过采样技术（Synthetic Minority Over-sampling Technique）。SMOTE 本质上是复制低频率分类的数据（横向或纵向翻转图像、改变饱和度等），以此来创建一个平衡的数据集。

8.4.1 其他医疗数据挑战

可以理解的是，医疗数据不像其他数据集那么容易发布，因此公共领域的医疗数据集更少。不过，这一现状正在慢慢变化，与此同时，你也可以尝试其他一些数据集和相关的挑战。需要注意的是，其中很多挑战已经被克服了，而幸运的是与这些挑战相关的数据集仍被继续发布。

8.4.2 ISBI 大挑战

ISBI（the IEEE International Symposium on Biomedical Imaging）是一个关于生物医学成

像的国际研讨会,这是一个很受欢迎的平台,它能够推进本章所涉及的相关工作。他们的年度会议常常以学术界面临的多项挑战为讨论重点。在 2016 年他们提出了几个挑战。

其中一个很受欢迎的挑战是内镜检查异常影像分析(Analysis of Images to Detect Abnormalities in Endoscopy,AIDA-E)。

另一个是淋巴结癌症转移检测(Cancer Metastasis Detection in Lymph Nodes,CMDLN),其特点是包含重要的病理数据。

在放射学方面,2016 年有一个广受欢迎的挑战是针对心脏疾病诊断的 Data Science Bowl 挑战。名称为 Transforming How We Diagnose Heart Disease 的挑战试图分割心脏磁共振成像数据的一部分以测量泵量,然后将其用作心脏健康的指标。

另一个流行的放射学数据集是:LIDC-IDRI 图像集合中的肺影像数据库联合的计算机断层扫描数据(Lung Image Database Consortium's Computed Tomography (CT) data)。这是一个肺癌筛查和诊断的胸部 CT 扫描数据集。有趣的是,该数据集并不是图像级别的分类数据集,而是对病变的实际位置作了注解。

这两个放射学竞赛之所以有趣,还有另外两个原因。

- ◆ 它们的特点是三维容积数据,这些数据本质上是一系列有序堆叠的二维图像,这些图像构成了实际的空间。
- ◆ 它们以分割任务为主要特点,你可以将其中一张图像或容积的某些部分分类到特定类别中,这是一种常见的分类挑战。除此之外,我们还尝试在图像中局部化特征。在一种情况下,我们试图局部化特征并指向它(而不是对整个图像进行分类);在另一种情况下,我们试图分类一个区域以度量区域大小。

稍后我们会讨论更多关于处理容积数据的内容。现在,你已经得到了一些可供使用的、不同的、有趣的数据集。

8.4.3 读取医疗数据

尽管糖尿病视网膜病变检测挑战赛具有一定的挑战性,但也不像它所表现的那么复杂。实际的图像是以 JPEG 格式提供的,但是大多数医疗数据并不是 JPEG 格式,它们使用诸如 DICOM 之类的格式。DICOM 代表医学数字成像和通信(Digital Imaging and Communications in Medicine),而且具有很多版本和变种。DICOM 数据不仅包含医疗图像,而且还包含头数据。头数据通常包括一般的人口统计学和研究数据,但也可以包含很多其他自定义的字段。有时头数据也会包含诊断信息,这些诊断信息可以用作标签。

DICOM 数据为之前讨论的管道添加了另一个步骤，即读取 DICOM 文件，提取头数据（可能是分类或标签数据），并提取潜在的图像。处理 DICOM 并不像处理 JPEG 或 PNG 那样容易，但也不是太困难。它需要一些额外的包。

因为我们几乎使用 Python 编写所有的东西，所以此处将使用一个 Python 库来处理 DICOM。处理 DICOM 最受欢迎的 Python 库是 pydicom。

应该注意的是，当前 pip 安装已经被破坏，所以必须从源代码库复制它，并通过设置脚本进行安装，然后才可以使用 pip。

文档中的快速摘要有助于理解如何处理 DICOM 文件。

```
>>> import dicom
>>> plan = dicom.read_file("rtplan.dcm")
>>> plan.PatientName
'Last^First^mid^pre'
>>> plan.dir("setup")    # get a list of tags with "setup"
somewhere in the name
['PatientSetupSequence']
>>> plan.PatientSetupSequence[0]
(0018, 5100) Patient Position                CS: 'HFS'
(300a, 0182) Patient Setup Number            IS: '1'
(300a, 01b2) Setup Technique Description     ST: ''
```

这可能看起来有点乱，但这就是在处理医疗数据时你应该期望的交互类型。更糟糕的是，即使是基本数据，每个供应商通常会对相同的数据设置稍有不同的标签。典型的行业惯例就是多看多了解。我们通过转储整个标签集来了解数据，如下所示。

```
>>> ds
(0008, 0012) Instance Creation Date     DA: '20030903'
(0008, 0013) Instance Creation Time     TM: '150031'
(0008, 0016) SOP Class UID              UI: RT Plan
Storage
(0008, 0018) Diagnosis                  UI: Positive
(0008, 0020) Study Date                 DA: '20030716'
(0008, 0030) Study Time                 TM: '153557'
(0008, 0050) Accession Number           SH: ''
(0008, 0060) Modality                   CS: 'RTPLAN'
```

假设我们在寻求诊断。我们可能会查看好几个标签文件，并尝试查看诊断结果是否持续在标签(0008，0018)诊断中出现，如果是，我们会将该字段从训练集中取出来，以查看它是否确实是一直增加的，以此来验证我们的假设。如果是一直增加的，那么就准备进行下一步；如果不是，那么就需要重新开始并查看其他字段。从理论上讲，数据提供者、代理商或供应商可以提供这些信息，但实际上并不是这么简单。

下一步是查看值的域。这一点很重要，因为我们想知道分类是什么样的。在理想情况下，将会有一组干净、漂亮的值的集合，例如{负值,正值}。但是实际上，我们经常会得到一个脏值长尾。因此，典型的方法是对每一个图像进行循环，并对遇到的每个单独的域值进行计数，如下所示。

```
>>> import dicom, glob, os
>>> os.chdir("/some/medical/data/dir")
>>> domains={}
>>> for file in glob.glob("*.dcm"):
>>>     aMedFile = dicom.read_file(file)
>>>     theVal=aMedFile.ds[0x10,0x10].value
>>>     if domains[theVal]>0:
>>>         domains[theVal]= domains[theVal]+1
>>>     else:
>>>         domains[theVal]=1
```

通常可以发现99%的域值都分布在少数的几个域值（例如正值和负值）上，并且有一个长尾为1%的域值是脏的（例如正值，但预览时显示@#Q#$%@#$%，或者发送以重新读取）。最简单的方法就是丢掉长尾，仅保留良好的数据。当存在大量训练数据时，这一做法特别容易。

虽然我们提取了分类信息，但是仍然需要提取实际的图像，操作如下。

```
>>> import dicom
>>> ds=dicom.read_file("MR_small.dcm")
>>> ds.pixel_array
array([[ 905, 1019, 1227, ...,  302,  304,  328],
       [ 628,  770,  907, ...,  298,  331,  355],
       [ 498,  566,  706, ...,  280,  285,  320],
       ...,
       [ 334,  400,  431, ..., 1094, 1068, 1083],
       [ 339,  377,  413, ..., 1318, 1346, 1336],
       [ 378,  374,  422, ..., 1369, 1129,  862]], dtype=int16)
>>> ds.pixel_array.shape
(64, 64)
```

但这只会得到一个原始的像素值矩阵。我们仍然需要将其转换成一种可读的格式（理想情况是 JPEG 或 PNG 格式）。我们将通过图 8-9 所示的代码完成下一步工作。

```
fileInputDICOM = r.rpop(redisQueue)
newFile = fileInputDICOM.replace(IN_DIR, OUT_DIR) + ".png"
print(num, ":", fileInputDICOM)
plan = dicomio.read_file(fileInputDICOM)
shape = plan.pixel_array.shape
wBuffer=MAX_SIZE-shape[0]
hBuffer=MAX_SIZE-shape[1]
image_2d = []
for row in plan.pixel_array:
    pixels = []
    for col in row:
        pixels.append(col)
    for h in range(hBuffer):
        pixels.append(32767)
    image_2d.append(pixels)
for w in range(wBuffer):
    image_2d.append([32767]*MAX_SIZE)

# Rescalling greyscale between 0-255
image_2d_scaled = []
for row in image_2d:
    row_scaled = []
    for col in row:
        col_scaled = int((float(col)/float(max_val))*255.0)
        col_scaled = 255.0 - col_scaled
        row_scaled.append(col_scaled)
    image_2d_scaled.append(row_scaled)

if not os.path.exists(os.path.dirname(newFile)):
    try:
        os.makedirs(os.path.dirname(newFile))
    except OSError as exc: # Guard against race condition
        if exc.errno != errno.EEXIST:
            raise

f = open(newFile, 'wb')
w = png.Writer(MAX_SIZE, MAX_SIZE, greyscale=True)
w.write(f, image_2d_scaled)
f.close()
```

图 8-9

接下来，把图像缩放到期望的位长，并利用另一个库将该矩阵写入一个文件中，该库可用我们的目标格式来写入数据。本例将使用 PNG 输出格式，并使用 png 库来写入。这意味着需要一些额外的导入。

```
import os
from pydicom import dicomio
import png
import errno
import fnmatch
```

我们将使用图 8-10 所示的代码来实现导出功能。

```
fileInputDICOM = r.rpop(redisQueue)
newFile = fileInputDICOM.replace(IN_DIR, OUT_DIR) + ".png"
print(num, ":", fileInputDICOM)
plan = dicomio.read_file(fileInputDICOM)
shape = plan.pixel_array.shape
wBuffer=MAX_SIZE-shape[0]
hBuffer=MAX_SIZE-shape[1]
image_2d = []
for row in plan.pixel_array:
    pixels = []
    for col in row:
        pixels.append(col)
    for h in range(hBuffer):
        pixels.append(32767)
    image_2d.append(pixels)
for w in range(wBuffer):
    image_2d.append([32767]*MAX_SIZE)

# Rescalling greyscale between 0-255
image_2d_scaled = []
for row in image_2d:
    row_scaled = []
    for col in row:
        col_scaled = int((float(col)/float(max_val))*255.0)
        col_scaled = 255.0 - col_scaled
        row_scaled.append(col_scaled)
    image_2d_scaled.append(row_scaled)

if not os.path.exists(os.path.dirname(newFile)):
    try:
        os.makedirs(os.path.dirname(newFile))
    except OSError as exc: # Guard against race condition
        if exc.errno != errno.EEXIST:
            raise

f = open(newFile, 'wb')
w = png.Writer(MAX_SIZE, MAX_SIZE, greyscale=True)
w.write(f, image_2d_scaled)
f.close()
```

图 8-10

8.5 所学技能

在本章中，你应该学到了这些技能。

◆ 处理晦涩难懂、独特的医疗影像格式。

◆ 处理庞大的图像文件，这（即庞大）是一个常见的医疗影像特点。

◆ 从医疗文件中提取分类数据。

- 扩展已有的管道，以处理异构数据输入。
- 应用非医疗数据预训练的网络。
- 扩展训练，以适应新的数据集。

8.6 总结

在本章中，我们创建了一个深度神经网络，并将其用于处理一个企业级的图像分类问题——医学诊断。此外，我们还引导学习了 DICOM 数字医学图像数据的读取，以进行深入的研究。在接下来的一章中，我们将创建一个可以通过学习用户的反馈来自我改进的生产系统。

第 9 章
生产系统自动化

在本章中,我们将创建一个生产系统,该系统能够区分 37 种不同种类的宠物(包括狗和猫)。用户可以将图片上传到该系统并接收识别结果。同时,该系统也可以接收用户的反馈,并每天自动进行自我训练以改善结果。

本章将关注以下几个方面。

- ◆ 如何将迁移学习应用到新的数据集。
- ◆ 如何将 TensorFlow Serving 应用于生产系统。
- ◆ 利用数据集的人群来源标签创建一个用户数据系统,并基于用户数据自动化地微调模型。

9.1 系统概述

系统的概览图如图 9-1 所示。

在该系统中,我们将在训练服务器上使用一份初始数据集来训练一个卷积神经网络模型。然后,该模型将在具有 TensorFlow Serving 功能的生产服务器上提供服务。生产服务器将会运行一个 Flask 服务器程序,它允许用户上传一张新图片,然后返回识别结果;如果模型给出了错误的结果,它还允许用户纠正标签。在每天固定的时间里,训练服务器会把用户标记的所有图像与当前数据集组合起来,自动微调模型并将其发送到生产服务器。Web 界面的线框图如图 9-2 所示,它允许用户上传图片和接收结果。

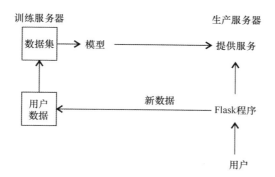

图 9-1

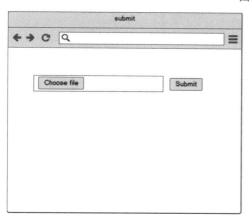 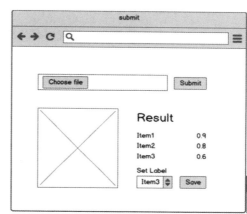

图 9-2

9.2 创建项目

在本章中,我们将对一个 VGG 模型进行微调,它已经在包含 1 000 个分类的 ImageNet 数据集上进行了训练。我们提供了一个包含预训练 VGG 模型和一些功能文件的初始项目,读者可以访问异步社区来获取代码。

在文件夹 chapter_09 中,你可以看到以下结构。

```
- data
--VGG16.npz
- samples_data
- production
- utils
--__init__.py
```

```
--debug_print.py
- README.md
```

你需要理解以下几点，其中包括两个文件的说明。

- VGG16.npz 是从 Caffe 模型导出的预训练模型。第 11 章将展示如何从 Caffe 模型创建该文件。在本章中，我们将其用作模型的初始值。你可以从文件夹 chapter_09 里的 README.md 下载该文件。

- production 是我们创建的一个 Flask 服务器，它提供一个 Web 界面以供用户上传图片并纠正模型。

- debug_print.py 包含了一些方法，本章将使用这些方法理解网络结构。

- samples_data 包含了一些猫、狗和汽车的图像，本章将使用这些图像。

9.3 加载预训练模型以加速训练

本节将重点关注在 TensorFlow 中加载预训练模型。我们将使用 VGG16 模型，该模型由牛津大学的 K. Simonyan 和 A. Zisserman 提出。

VGG16 是一个非常深的神经网络，包含了大量的卷积层、最大池化层和全连接层，如图 9-3 所示。在 ImageNet 挑战中，针对一个包含 1000 种图像分类的验证集，VGG16 模型在单尺度方法排名前 5 的分类误差为 8.1%。

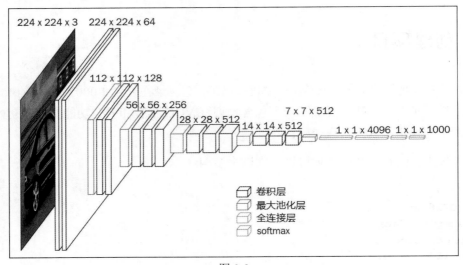

图 9-3

首先，在项目目录中创建一个名为 nets.py 的文件。下面代码定义了 VGG16 模型的计算图。

```python
import tensorflow as tf
import numpy as np


def inference(images):
    with tf.name_scope("preprocess"):
        mean = tf.constant([123.68, 116.779, 103.939], dtype=tf.float32, shape=[1, 1, 1, 3], name='img_mean')
        input_images = images - mean
    conv1_1 = _conv2d(input_images, 3, 3, 64, 1, 1, name="conv1_1")
    conv1_2 = _conv2d(conv1_1, 3, 3, 64, 1, 1, name="conv1_2")
    pool1 = _max_pool(conv1_2, 2, 2, 2, 2, name="pool1")

    conv2_1 = _conv2d(pool1, 3, 3, 128, 1, 1, name="conv2_1")
    conv2_2 = _conv2d(conv2_1, 3, 3, 128, 1, 1, name="conv2_2")
    pool2 = _max_pool(conv2_2, 2, 2, 2, 2, name="pool2")

    conv3_1 = _conv2d(pool2, 3, 3, 256, 1, 1, name="conv3_1")
    conv3_2 = _conv2d(conv3_1, 3, 3, 256, 1, 1, name="conv3_2")
    conv3_3 = _conv2d(conv3_2, 3, 3, 256, 1, 1, name="conv3_3")
    pool3 = _max_pool(conv3_3, 2, 2, 2, 2, name="pool3")

    conv4_1 = _conv2d(pool3, 3, 3, 512, 1, 1, name="conv4_1")
    conv4_2 = _conv2d(conv4_1, 3, 3, 512, 1, 1, name="conv4_2")
    conv4_3 = _conv2d(conv4_2, 3, 3, 512, 1, 1, name="conv4_3")
    pool4 = _max_pool(conv4_3, 2, 2, 2, 2, name="pool4")

    conv5_1 = _conv2d(pool4, 3, 3, 512, 1, 1, name="conv5_1")
    conv5_2 = _conv2d(conv5_1, 3, 3, 512, 1, 1, name="conv5_2")
    conv5_3 = _conv2d(conv5_2, 3, 3, 512, 1, 1, name="conv5_3")
    pool5 = _max_pool(conv5_3, 2, 2, 2, 2, name="pool5")

    fc6 = _fully_connected(pool5, 4096, name="fc6")
    fc7 = _fully_connected(fc6, 4096, name="fc7")
    fc8 = _fully_connected(fc7, 1000, name='fc8', relu=False)
    outputs = _softmax(fc8, name="output")
    return outputs
```

在上述代码中，需要注意以下事项。

- _conv2d、_max_pool、_fully_connected 和 _softmax 方法分别定义了卷积层、最大池化层、全连接层和 softmax 层，稍后将实现这些方法。
- 在命名空间 preprocess 中定义一个常量张量 mean，输入图像都会减去该张量值。得到的结果就是训练 VGG16 模型使得图像均值为零的均值向量。
- 然后，使用参数定义卷积层、最大池化层和全连接层。
- 在 fc8 层中，我们未将 ReLU 激励应用到输出上，而是将输出传递给一个 softmax 层，以计算在 1 000 多个分类上的概率。

现在，在 nets.py 文件中实现 _conv2d、_max_pool、_fully_connected 和 _softmax。

下面是方法 _conv2d 和 _max_pool 的实现代码。

```
def _conv2d(input_data, k_h, k_w, c_o, s_h, s_w, name, relu=True, padding="SAME"):
    c_i = input_data.get_shape()[-1].value
    convolve = lambda i, k: tf.nn.conv2d(i, k, [1, s_h, s_w, 1], padding=padding)
    with tf.variable_scope(name) as scope:
        weights = tf.get_variable(name="kernel", shape=[k_h, k_w, c_i, c_o], initializer=tf.truncated_normal_initializer(stddev=1e-1, dtype=tf.float32))
        conv = convolve(input_data, weights)
        biases = tf.get_variable(name="bias", shape=[c_o], dtype=tf.float32, initializer=tf.constant_initializer(value=0.0))
        output = tf.nn.bias_add(conv, biases)
        if relu:
            output = tf.nn.relu(output, name=scope.name)
        return output
def _max_pool(input_data, k_h, k_w, s_h, s_w, name, padding="SAME"):
    return tf.nn.max_pool(input_data, ksize=[1, k_h, k_w, 1],
                          strides=[1, s_h, s_w, 1],
 padding=padding,
 name=name)
```

如果你已经阅读过第 4 章，那么会发现上述大部分代码之前讲解过，但是有几行需要解释一下。

- k_h 和 k_w 是内核的高度和权重。
- c_o 表示通道输出,它是卷积层特征映射的数量。
- s_h 和 s_w 是 `tf.nn.conv2d` 层和 `tf.nn.max_pool` 层的步幅参数。
- 这里使用了 `tf.get_variable` 而非 `tf.Variable`,是因为加载预训练权重时需要再次使用 `get_variable`。

实现全连接层和 softmax 层的代码如下。

```
def _fully_connected(input_data, num_output, name, relu=True):
    with tf.variable_scope(name) as scope:
        input_shape = input_data.get_shape()
        if input_shape.ndims == 4:
            dim = 1
            for d in input_shape[1:].as_list():
                dim *= d
            feed_in = tf.reshape(input_data, [-1, dim])
        else:
            feed_in, dim = (input_data, input_shape[-1].value)
        weights = tf.get_variable(name="kernel", shape=[dim, num_output], initializer=tf.truncated_normal_initializer(stddev=1e-1, dtype=tf.float32))
        biases = tf.get_variable(name="bias", shape=[num_output], dtype=tf.float32, initializer=tf.constant_initializer(value=0.0))
        op = tf.nn.relu_layer if relu else tf.nn.xw_plus_b
        output = op(feed_in, weights, biases, name=scope.name)
        return output
def _softmax(input_data, name):
    return tf.nn.softmax(input_data, name=name)
```

利用 `_fully_connected` 方法,我们首先检查输入数据的维数,以便将输入数据重新调整为正确的速度。然后,利用 `get_variable` 方法创建变量 `weights` 和 `biases`。最后,检查参数 `relu`,以决定是否利用 `tf.nn.relu_layer` 或 `tf.nn.xw_plus_b` 将 Relu 应用到输出上。`tf.nn.relu_layer` 将计算 `relu(matmul(x, weights) + biases)`,而 `tf.nn.xw_plus_b` 则只会计算 `matmul(x, weights) + biases`。

本节的最后一个方法用于将预训练的 Caffe 权重加载到定义的变量中。

```
def load_caffe_weights(path, sess, ignore_missing=False):
    print("Load caffe weights from ", path)
```

```
        data_dict = np.load(path).item()
        for op_name in data_dict:
            with tf.variable_scope(op_name, reuse=True):
                for param_name, data in
data_dict[op_name].iteritems():
                    try:
                        var = tf.get_variable(param_name)
                        sess.run(var.assign(data))
                    except ValueError as e:
                        if not ignore_missing:
                            print(e)
                            raise e
```

为了理解该方法，我们必须知道数据在预训练模型 VGG16.npz 中是如何存储的。我们已经创建了一段简单的代码，以此来输出预训练模型中的所有变量。现在可以将下面代码存放在 nets.py 的结尾处，并用命令 python nets.py 运行它。

```
if __name__ == "__main__":
path = "data/VGG16.npz"
data_dict = np.load(path).item()
for op_name in data_dict:
    print(op_name)
    for param_name, data in data_dict[op_name].iteritems():
        print("\t" + param_name + "\t" + str(data.shape))
```

下面是部分输出结果。

```
conv1_1
    weights   (3, 3, 3, 64)
    biases    (64,)
conv1_2
    weights   (3, 3, 64, 64)
    biases    (64,)
```

正如你所看到的，`op_name` 是网络层的名称，并且能通过 `data_dict[op_name]` 访问每一层的权重和偏差。

下面看一下 `load_caffe_weights`。

- 在参数中使用它配合 `tf.variable_scope` 和 `reuse=True`，可以得到在计算图中定义的 `weights` 和 `biases` 的确切变量。在此之后，运行分配方法来为每个变量设置数据。

- 如果变量名未定义，那么方法 `get_variable` 将会给出 `ValueError`。因此，我们将使用变量 ignore_missing 来决定是否应该抛出错误。

测试预训练模型

前面已经创建了一个 VGG16 神经网络。在此，我们将尝试使用预训练的模型来分类汽车、猫和狗，以检查是否成功地加载了模型。

在 nets.py 文件中，需要使用以下代码替换当前的 __main__ 代码。

```python
import os
from utils import debug_print
from scipy.misc import imread, imresize


if __name__ == "__main__":
 SAMPLES_FOLDER = "samples_data"
 with open('%s/imagenet-classes.txt' % SAMPLES_FOLDER, 'rb') as infile:
  class_labels = map(str.strip, infile.readlines())

 inputs = tf.placeholder(tf.float32, [None, 224, 224, 3], name="inputs")
 outputs = inference(inputs)

 debug_print.print_variables(tf.global_variables())
 debug_print.print_variables([inputs, outputs])

 with tf.Session() as sess:
  load_caffe_weights("data/VGG16.npz", sess, ignore_missing=False)

     files = os.listdir(SAMPLES_FOLDER)
     for file_name in files:
         if not file_name.endswith(".jpg"):
             continue
         print("=== Predict %s ==== " % file_name)
         img = imread(os.path.join(SAMPLES_FOLDER, file_name), mode="RGB")
         img = imresize(img, (224, 224))

         prob = sess.run(outputs, feed_dict={inputs: [img]})[0]
         preds = (np.argsort(prob)[::-1])[0:3]
```

```
        for p in preds:
            print class_labels[p], prob[p]
```

在上述代码中,有以下几点需要注意。

- 利用 `debug_print.print_variables` 方法,我们通过输出变量名和维度来可视化所有的变量。
- 定义一个形状为 `[None, 224, 224, 3]` 的占位符 `inputs`,这是 VGG16 模型所需要的输入尺寸。

  We get the model graph with outputs = inference(inputs).

- 在 `tf.Session()` 中,用 `ignore_missing=False` 选项调用 `load_caffe_weights` 方法,以确保可以加载预训练模型所有的权重和偏差。
- 利用 `scipy` 包中的 `imread` 和 `imresize` 方法分别加载和调整图像。然后,利用字典 `feed_dict` 使用方法 `sess.run` 并接收预测值。
- 下面的结果是 samples_data 中的 car.jpg、cat.jpg 和 dog.jpg 的预测值,其中 samples_data 是本章开头处所提供的数据。

```
== Predict car.jpg ====
racer, race car, racing car 0.666172
sports car, sport car 0.315847
car wheel 0.0117961
=== Predict cat.jpg ====
Persian cat 0.762223
tabby, tabby cat 0.0647032
lynx, catamount 0.0371023
=== Predict dog.jpg ====
Border collie 0.562288
collie 0.239735
Appenzeller 0.0186233
```

上述结果是这些图像的真实标签。这意味着我们已经在 TensorFlow 中成功地加载了预训练的 VGG16 模型。在下一节中,我们将展示如何在数据集上对模型进行微调。

9.4 为数据集训练模型

在本节中,我们将完成创建数据集、微调模型和导出用于生产的模型等任务。

9.4.1 Oxford-IIIT 宠物数据集介绍

Oxford-IIIT 宠物数据集（Oxford-IIIT pet dataset）包含 37 个种类的狗和猫，每个种类都包含大约 200 张在比例、姿势和亮度上有很大不同的图片。真实值数据为每张图片提供了种类注解、头部位置和像素分割等信息，如图 9-4 所示。在我们的应用程序中，仅仅使用物种名称作为模型的分类名。

图 9-4

1．数据集统计信息

以下是狗和猫的品种的数据集的情况。

（1）狗的品种如表 9-1 所示。

表 9-1

品　　种	总数（张）
美国斗牛犬	200
美国比特犬	200
矮腿猎犬	200
比格犬	200
拳师犬	199
吉娃娃犬	200
英国可卡犬	196
英国赛特犬	200
德国短毛波音达犬	200
大白熊犬	200
哈威那	200
日本狆	200
荷兰毛狮犬	199
兰波格犬	200
迷你杜宾犬	200

续表

品　　种	总数（张）
纽芬兰犬	196
博美犬	200
巴哥犬	200
圣伯纳犬	200
萨摩耶犬	200
苏格兰梗	199
日本柴犬	200
斯塔福郡斗牛梗	189
爱尔兰软毛梗	200
约克夏梗犬	200
总数	4978

（2）猫的品种如表 9-2 所示。

表 9-2

品　　种	总数（张）
阿比西尼亚猫	198
孟加拉豹猫	200
巴曼猫	200
孟买猫	184
英国短毛猫	200
埃及猫	190
缅因库恩猫	200
波斯猫	200
布偶猫	200
俄罗斯蓝猫	200
暹罗猫	199
加拿大无毛猫	200
总数	2371

（3）宠物总量如表 9-3 所示。

表 9-3

科	数量（张）
猫	2371
狗	4978
总数	7349

2．下载数据集

可以从牛津大学网站获取数据集。我们需要下载图像数据集 images.tar.gz 和真实值数据 annotations.tar.gz，将 TAR 文件存储在 data/datasets 文件夹中，并提取所有的.tar 文件。请确保 data 文件夹具有以下结构。

```
- data
-- VGG16.npz
-- datasets
---- annotations
------ trainval.txt
---- images
------ *.jpg
```

3．准备数据

在开始训练之前，需要将数据集预处理成一种更简单的格式，以用于进一步的自动化微调。

首先，在 project 文件夹中创建一个名为 scripts 的 Python 包。然后，创建一个名为 convert_oxford_data.py 的 Python 文件，并添加以下代码。

```python
import os
import tensorflow as tf
from tqdm import tqdm
from scipy.misc import imread, imsave

FLAGS = tf.app.flags.FLAGS

tf.app.flags.DEFINE_string(
'dataset_dir', 'data/datasets',
'The location of Oxford IIIT Pet Dataset which contains
 annotations and images folders'
)
```

```python
tf.app.flags.DEFINE_string(
'target_dir', 'data/train_data',
'The location where all the images will be stored'
)

def ensure_folder_exists(folder_path):
    if not os.path.exists(folder_path):
        os.mkdir(folder_path)
    return folder_path

def read_image(image_path):
    try:
        image = imread(image_path)
        return image
    except IOError:
        print(image_path, "not readable")
    return None
```

在这段代码中,我们使用 `tf.app.flags.FLAGS` 解析参数,以便轻易地自定义脚本。另外,我们还创建了两个辅助方法来创建目录和读取图像。

接下来,添加以下代码,将 Oxford-IIIT 数据集转换成所需要的格式。

```python
def convert_data(split_name, save_label=False):
    if split_name not in ["trainval", "test"]:
        raise ValueError("split_name is not recognized!")
    target_split_path = ensure_folder_exists(os.path.join(FLAGS.target_dir, split_name))
    output_file = open(os.path.join(FLAGS.target_dir, split_name + ".txt"), "w")

    image_folder = os.path.join(FLAGS.dataset_dir, "images")
    anno_folder = os.path.join(FLAGS.dataset_dir, "annotations")

    list_data = [line.strip() for line in open(anno_folder + "/" + split_name + ".txt")]

    class_name_idx_map = dict()
    for data in tqdm(list_data, desc=split_name):
        file_name, class_index, species, breed_id = data.split(" ")
        file_label = int(class_index) - 1

        class_name = "_".join(file_name.split("_")[0:-1])
```

```
            class_name_idx_map[class_name] = file_label

            image_path = os.path.join(image_folder, file_name + ".jpg")
            image = read_image(image_path)
            if image is not None:
            target_class_dir =
             ensure_folder_exists(os.path.join(target_split_path,
             class_name))
            target_image_path = os.path.join(target_class_dir,
             file_name + ".jpg")
                imsave(target_image_path, image)
                output_file.write("%s %s\n" % (file_label,
                target_image_path))

    if save_label:
        label_file = open(os.path.join(FLAGS.target_dir,
        "labels.txt"), "w")
        for class_name in sorted(class_name_idx_map,
        key=class_name_idx_map.get):
        label_file.write("%s\n" % class_name)


def main(_):
    if not FLAGS.dataset_dir:
    raise ValueError("You must supply the dataset directory with
    --dataset_dir")

    ensure_folder_exists(FLAGS.target_dir)
    convert_data("trainval", save_label=True)
    convert_data("test")


if __name__ == "__main__":
    tf.app.run()
```

现在，使用以下代码来运行脚本。

python scripts/convert_oxford_data.py --dataset_dir data/datasets/ --target_dir data/train_data

该脚本读取了 Oxford-IIIT 数据集的真实值数据，并以下结构在 data/train_data 目录中创建了一个新的数据集。

- **train_data**

```
-- trainval.txt
-- test.txt
-- labels.txt
-- trainval
---- Abyssinian
---- ...
-- test
---- Abyssinian
---- ...
```

下面是对数据集文件的解释。

◆ labels.txt 包含了数据集中的 37 个宠物种类。

◆ trainval.txt 包含了在训练过程中使用的图片的列表，格式为<class_id> <image_path>。

◆ test.txt 包含了将要用来检查模型准确性的图片的列表。test.txt 的格式与 trainval.txt 的格式相同。

◆ trainval 文件夹和 test 文件夹各包含 37 个子文件夹，每个子文件夹都对应每个宠物种类的名称，并包含每个种类的所有图片。

9.4.2　为训练和测试创建输入管道

TensorFlow 允许创建可靠的输入管道，以进行快速而简便的训练。在本节中，我们将使用 `tf.TextLineReader` 来读取训练文本文件和测试文本文件，并使用 `tf.train.batch` 并行地读取和预处理图像。

首先，需要在 project 目录中创建一个新的 Python 文件 datasets.py，并添加以下代码。

```
import tensorflow as tf
import os

def load_files(filenames):
  filename_queue = tf.train.string_input_producer(filenames)
  line_reader = tf.TextLineReader()
  key, line = line_reader.read(filename_queue)
  label, image_path = tf.decode_csv(records=line,
    record_defaults=[tf.constant([], dtype=tf.int32),
    tf.constant([], dtype=tf.string)],
                                    field_delim=' ')
  file_contents = tf.read_file(image_path)
  image = tf.image.decode_jpeg(file_contents, channels=3)
```

```
    return image, label
```

在 `load_files` 方法中，使用 `tf.TextLineReader` 读取文本文件（如 trainval.txt 和 test.txt）的每一行。`tf.TextLineReader` 需要读取一个字符串队列，我们可以使用 `tf.train.string_input_producer` 来存储文件名。在此之后，将行变量传送到 `tf.decode_cvs` 中，以获得标签和文件名。此外，利用 `tf.image.decode_jpeg` 可以轻易地读取图像。

既然现在可以加载图像了，那么我们将继续进行下一步操作，为训练工作创建图像批次和标签批次。

在 datasets.py 中添加一个新方法。

```
def input_pipeline(dataset_dir, batch_size, num_threads=8,
   is_training=True, shuffle=True):
   if is_training:
       file_names = [os.path.join(dataset_dir, "trainval.txt")]
   else:
       file_names = [os.path.join(dataset_dir, "test.txt")]
   image, label = load_files(file_names)

   image = preprocessing(image, is_training)

   min_after_dequeue = 1000
   capacity = min_after_dequeue + 3 * batch_size
   if shuffle:
    image_batch, label_batch = tf.train.shuffle_batch(
    [image, label], batch_size, capacity,
    min_after_dequeue, num_threads
     )
   else:
       image_batch, label_batch = tf.train.batch(
           [image, label], batch_size, num_threads, capacity
           )
   return image_batch, label_batch
```

首先，使用 `load_files` 方法加载图像和标签。然后，通过一个新的预处理方法来传送图像，我们稍后将实现该方法。最后，将图像和标签分别传送到 `tf.train.shuffle_batch` 和 `tf.train.batch` 中，以供训练和测试。

```
def preprocessing(image, is_training=True, image_size=224,
  resize_side_min=256, resize_side_max=312):
    image = tf.cast(image, tf.float32)
```

```
    if is_training:
        resize_side = tf.random_uniform([], minval=resize_side_min,
        maxval=resize_side_max+1, dtype=tf.int32)
        resized_image = _aspect_preserving_resize(image,
        resize_side)

        distorted_image = tf.random_crop(resized_image,
 [image_size,
        image_size, 3])

        distorted_image =
        tf.image.random_flip_left_right(distorted_image)
        distorted_image =
        tf.image.random_brightness(distorted_image, max_delta=50)

        distorted_image = tf.image.random_contrast(distorted_image,
        lower=0.2, upper=2.0)

        return distorted_image
    else:
        resized_image = _aspect_preserving_resize(image,
 image_size)
        return
 tf.image.resize_image_with_crop_or_pad(resized_image,
        image_size, image_size)
```

在训练和测试过程中我们有两种不同的方法来对数据进行预处理。在训练过程中，我们需要为当前数据集增加数据，以创建更多训练数据。下面是预处理方法用到的一些技术。

- 数据集中的图片可能具有不同的图像分辨率，而我们只需要分辨率为 224×224 的图像。在执行随机裁剪之前，需要将图像调整为一个合理的大小。图 9-5 描述了如何对图像进行裁剪。

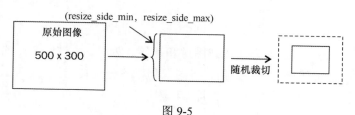

图 9-5

- 裁剪完图像后，可以通过 tf.image.random_flip_left_right、tf.image.random_brightness 和 tf.image.random_contrast 来调整图像并创建一

个新的训练样本。

◆ 在测试流程中，只需要利用_aspect_preserving_resize 和 tf.image.resize_image_with_crop_or_pad 来调整图像的大小即可。tf.image.resize_image_with_crop_or_pad 允许我们在图像中心裁剪，或者将图像填充到目标宽度和高度。

现在，需要将最后两个方法添加到 datasets.py 中，如下所示。

```
def _smallest_size_at_least(height, width, smallest_side):
  smallest_side = tf.convert_to_tensor(smallest_side,
  dtype=tf.int32)

  height = tf.to_float(height)
  width = tf.to_float(width)
  smallest_side = tf.to_float(smallest_side)

  scale = tf.cond(tf.greater(height, width),
            lambda: smallest_side / width,
            lambda: smallest_side / height)
  new_height = tf.to_int32(height * scale)
  new_width = tf.to_int32(width * scale)
  return new_height, new_width

def _aspect_preserving_resize(image, smallest_side):
  smallest_side = tf.convert_to_tensor(smallest_side,
  dtype=tf.int32)
  shape = tf.shape(image)
  height = shape[0]
  width = shape[1]
  new_height, new_width = _smallest_size_at_least(height, width,
  smallest_side)
  image = tf.expand_dims(image, 0)
  resized_image = tf.image.resize_bilinear(image, [new_height,
  new_width], align_corners=False)
  resized_image = tf.squeeze(resized_image)
  resized_image.set_shape([None, None, 3])
  return resized_image
```

到本节为止，我们已经做了大量的工作来准备数据集和输入管道。在下一节中，我们将为数据集、损失值、准确率和训练操作等定义模型，以执行训练过程。

9.4.3 定义模型

我们的应用程序需要对 37 个种类的狗和猫进行分类，而 VGG16 模型支持 1 000 种不同的类别。在我们的应用程序中，我们将再次使用层（直到 fc7 层），并从头开始训练最后一层。为了使模型输出 37 个类别，需要在 nets.py 中修改 inference 方法，如下所示。

```
def inference(images, is_training=False):
#
# All the code before fc7 are not modified.
#
fc7 = _fully_connected(fc6, 4096, name="fc7")
if is_training:
    fc7 = tf.nn.dropout(fc7, keep_prob=0.5)
fc8 = _fully_connected(fc7, 37, name='fc8-pets', relu=False)
return fc8
```

- 在方法中添加一个新的参数 `is_training`。在 fc7 层后，如果 inference 方法用于训练，那么我们将添加一个 `tf.nn.dropout` 层，该层有助于模型在未知数据下进行更好的调整，以避免过度拟合。

- fc8 层输出的数量从 1 000 变成了 37。此外，必须将 fc8 层的名称更改为另一个名称：在本例中，将其修改为 fc8-pets。如果不改变 fc8 层的名称，那么 `load_caffe_weights` 仍然会找到新的网络层并分配原始的权重，但它与新的 fc8 层的尺寸并不相同。

- 移除 `inference` 方法末尾的 `softmax` 层，因为稍后使用的损失函数仅需要未标准化的输出。

9.4.4 定义训练操作

我们将在一个新的 Python 文件 models.py 中定义所有的操作。首先，创建一些操作来计算损失值和准确率。

```
def compute_loss(logits, labels):
  labels = tf.squeeze(tf.cast(labels, tf.int32))

  cross_entropy =
tf.nn.sparse_softmax_cross_entropy_with_logits(logits=logits,
labels=labels)
  cross_entropy_mean = tf.reduce_mean(cross_entropy)
  tf.add_to_collection('losses', cross_entropy_mean)
```

```
  return tf.add_n(tf.get_collection('losses'),
  name='total_loss')


def compute_accuracy(logits, labels):
  labels = tf.squeeze(tf.cast(labels, tf.int32))
  batch_predictions = tf.cast(tf.argmax(logits, 1), tf.int32)
  predicted_correctly = tf.equal(batch_predictions, labels)
  accuracy = tf.reduce_mean(tf.cast(predicted_correctly,
  tf.float32))
  return accuracy
```

在这些方法中，logits 是模型的输出，而 labels 是数据集中的真实值数据。compute_loss 方法使用了 tf.nn.sparse_softmax_cross_entropy_with_logits，从而不需要使用 softmax 方法标准化 logits。此外，不需要将 labels 转换成独热向量。compute_accuracy 方法将 logits 中的最大值与 tf.argmax 和 labels 进行比较，以获得准确率。

接下来，为 learning_rate 和 optimizer 定义操作。

```
def get_learning_rate(global_step, initial_value, decay_steps,
  decay_rate):
  learning_rate = tf.train.exponential_decay(initial_value,
  global_step, decay_steps, decay_rate, staircase=True)
  return learning_rate


def train(total_loss, learning_rate, global_step, train_vars):

  optimizer = tf.train.AdamOptimizer(learning_rate)

  train_variables = train_vars.split(",")

  grads = optimizer.compute_gradients(
      total_loss,
      [v for v in tf.trainable_variables() if v.name in
      train_variables]
      )
  train_op = optimizer.apply_gradients(grads,
  global_step=global_step)
  return train_op
```

在 `train` 方法中，将 `optimizer` 配置成仅对在 `train_vars` 字符串中定义的一些变量进行计算和应用梯度，这将允许我们仅更新最后一层即 fc8 层的权重和偏差，并冻结其他层。`train_vars` 是一个包含多个以逗号分隔的变量的字符串，例如 `models/fc8-pets/weights:0`、`models/fc8-pets/biases:0`。

9.4.5 执行训练过程

现在我们已经为训练模型做好了准备。接下来，在 scripts 文件夹中创建一个名为 train.py 的 Python 文件。首先，需要为训练过程定义一些参数。

```python
import tensorflow as tf
import os
from datetime import datetime
from tqdm import tqdm

import nets, models, datasets

# Dataset
dataset_dir = "data/train_data"
batch_size = 64
image_size = 224

# Learning rate
initial_learning_rate = 0.001
decay_steps = 250
decay_rate = 0.9

# Validation
output_steps = 10  # Number of steps to print output
eval_steps = 20  # Number of steps to perform evaluations

# Training
max_steps = 3000  # Number of steps to perform training
save_steps = 200  # Number of steps to perform saving checkpoints
num_tests = 5  # Number of times to test for test accuracy
max_checkpoints_to_keep = 3
save_dir = "data/checkpoints"
train_vars = 'models/fc8-pets/weights:0,models/fc8-pets/biases:0'

# Export
export_dir = "/tmp/export/"
export_name = "pet-model"
```

```
export_version = 2
```

这些变量都是浅显易懂的。接下来，需要为训练过程定义一些操作，如下所示。

```
images, labels = datasets.input_pipeline(dataset_dir, batch_size,
is_training=True)
test_images, test_labels = datasets.input_pipeline(dataset_dir,
batch_size, is_training=False)

with tf.variable_scope("models") as scope:
    logits = nets.inference(images, is_training=True)
    scope.reuse_variables()
    test_logits = nets.inference(test_images, is_training=False)

total_loss = models.compute_loss(logits, labels)
train_accuracy = models.compute_accuracy(logits, labels)
test_accuracy = models.compute_accuracy(test_logits, test_labels)
global_step = tf.Variable(0, trainable=False)
learning_rate = models.get_learning_rate(global_step,
initial_learning_rate, decay_steps, decay_rate)
train_op = models.train(total_loss, learning_rate, global_step,
train_vars)

saver = tf.train.Saver(max_to_keep=max_checkpoints_to_keep)
checkpoints_dir = os.path.join(save_dir,
datetime.now().strftime("%Y-%m-%d_%H-%M-%S"))
if not os.path.exists(save_dir):
    os.mkdir(save_dir)
if not os.path.exists(checkpoints_dir):
    os.mkdir(checkpoints_dir)
```

这些操作是通过调用在 datasets.py、nets.py 和 models.py 中定义的方法而实现的。在这段代码中，我们创建了一个用于训练的输入管道和一个用于测试的管道。在此之后，创建了一个名为 models 的新的 variable_scope，并利用方法 nets.inference 创建了 logits 和 test_logits。必须确保添加了 scope.reuse_variables，因为我们将在测试中再次使用训练时的权重和偏差。最后，创建了一个 saver 来保存每个 save_steps 的检查点。

训练过程的最后一部分是训练循环。

```
with tf.Session() as sess:
    sess.run(tf.global_variables_initializer())
    coords = tf.train.Coordinator()
```

```python
        threads = tf.train.start_queue_runners(sess=sess, coord=coords)

        with tf.variable_scope("models"):
            nets.load_caffe_weights("data/VGG16.npz", sess,
            ignore_missing=True)

        last_saved_test_accuracy = 0
        for i in tqdm(range(max_steps), desc="training"):
                _, loss_value, lr_value = sess.run([train_op,
                    total_loss, learning_rate])

    if (i + 1) % output_steps == 0:
        print("Steps {}: Loss = {:.5f} Learning Rate =
        {}".format(i + 1, loss_value, lr_value))

    if (i + 1) % eval_steps == 0:
        test_acc, train_acc, loss_value =
        sess.run([test_accuracy, train_accuracy, total_loss])
        print("Test accuracy {} Train accuracy {} : Loss =
        {:.5f}".format(test_acc, train_acc, loss_value))

    if (i + 1) % save_steps == 0 or i == max_steps - 1:
        test_acc = 0
        for i in range(num_tests):
            test_acc += sess.run(test_accuracy)
        test_acc /= num_tests
    if test_acc > last_saved_test_accuracy:
            print("Save steps: Test Accuracy {} is higher than
            {}".format(test_acc, last_saved_test_accuracy))
            last_saved_test_accuracy = test_acc
            saved_file = saver.save(sess,
    os.path.join(checkpoints_dir, 'model.ckpt'),
                global_step=global_step)
            print("Save steps: Save to file %s " % saved_file)
        else:
            print("Save steps: Test Accuracy {} is not higher
                than {}".format(test_acc,
last_saved_test_accuracy))

        models.export_model(checkpoints_dir, export_dir, export_name,
        export_version)

        coords.request_stop()
```

```
coords.join(threads)
```

训练循环很容易理解。首先，因为之前更改了 fc8 层的名称，所以可以通过将 ignore_missing 设置为 True 来加载预训练的 VGG16 模型。然后，循环 max_steps 步，每 output_steps 步输出 loss，每 eval_steps 步输出 test_accuracy。如果当前测试准确率高于前一准确率，那么每 save_steps 步会检查并保存检查点。我们仍然需要创建 models.export_model 来导出模型，以供训练结束后使用。然而，在继续前进之前，可能需要检查一下训练流程是否正常工作。为此，将以下行注释。

```
models.export_model(checkpoints_dir, export_dir, export_name,
export_version)
```

然后，使用以下命令运行训练脚本。

```
python scripts/train.py
```

下面是控制台中的一些输出。脚本先加载了预训练的模型。然后，它将输出损失值。

```
('Load caffe weights from ', 'data/VGG16.npz')
training:    0%|                    | 9/3000 [00:05<24:59,  1.99it/s]
Steps 10: Loss = 31.10747 Learning Rate = 0.0010000000475
training:    1%|                    | 19/3000 [00:09<19:19,  2.57it/s]
Steps 20: Loss = 34.43741 Learning Rate = 0.0010000000475
Test accuracy 0.296875 Train accuracy 0.0 : Loss = 31.28600
training:    1%|                    | 29/3000 [00:14<20:01,  2.47it/s]
Steps 30: Loss = 15.81103 Learning Rate = 0.0010000000475
training:    1%|                    | 39/3000 [00:18<19:42,  2.50it/s]
Steps 40: Loss = 14.07709 Learning Rate = 0.0010000000475
Test accuracy 0.53125 Train accuracy 0.03125 : Loss = 20.65380
```

现在，停止训练，并取消对 export_model 方法的注释。使用 models.export_model 方法将具有最高测试准确率的最新模型导出到 export_dir 文件夹，模型的名称为 export_name，版本为 export_version。

9.4.6 导出模型以用于生产

导出模型的代码如下。

```
def export_model(checkpoint_dir, export_dir, export_name,
export_version):
    graph = tf.Graph()
    with graph.as_default():
        image = tf.placeholder(tf.float32, shape=[None, None, 3])
        processed_image = datasets.preprocessing(image,
```

```
    is_training=False)
with tf.variable_scope("models"):
 logits = nets.inference(images=processed_image,
   is_training=False)

model_checkpoint_path =
get_model_path_from_ckpt(checkpoint_dir)
saver = tf.train.Saver()

config = tf.ConfigProto()
config.gpu_options.allow_growth = True
config.gpu_options.per_process_gpu_memory_fraction = 0.7

with tf.Session(graph=graph) as sess:
    saver.restore(sess, model_checkpoint_path)
    export_path = os.path.join(export_dir, export_name,
    str(export_version))
    export_saved_model(sess, export_path, image, logits)
    print("Exported model at", export_path)
```

在 export_model 方法中,我们需要创建一个在生产中运行的新计算图。在生产中不像在训练中那样需要所有的变量,并且也不需要输入管道。但是,需要使用 export_saved_model 方法导出模型,如下所示。

```
def export_saved_model(sess, export_path, input_tensor,
output_tensor):
    from tensorflow.python.saved_model import builder as
saved_model_builder
    from tensorflow.python.saved_model import signature_constants
    from tensorflow.python.saved_model import signature_def_utils
    from tensorflow.python.saved_model import tag_constants
    from tensorflow.python.saved_model import utils
    builder = saved_model_builder.SavedModelBuilder(export_path)

    prediction_signature = signature_def_utils.build_signature_def(
        inputs={'images': utils.build_tensor_info(input_tensor)},
        outputs={
            'scores': utils.build_tensor_info(output_tensor)
        },
        method_name=signature_constants.PREDICT_METHOD_NAME)

    legacy_init_op = tf.group(
        tf.tables_initializer(), name='legacy_init_op')
```

```
builder.add_meta_graph_and_variables(
    sess, [tag_constants.SERVING],
    signature_def_map={
      'predict_images':
        prediction_signature,
    },
    legacy_init_op=legacy_init_op)

builder.save()
```

利用该方法，我们就可以为用于生产的模型创建一个元图（metagraph）。我们将在后续章节中讨论如何利用模型提供服务。现在，运行脚本进行自动化训练，并在 3 000 步后导出模型。

python scripts/train.py

在一个包含了核心 i7-4790 CPU 和一个 TITAN-X GPU 的系统上，训练过程花费了 20min。下面是控制台中最后的一些输出内容。

```
Steps 3000: Loss = 0.59160 Learning Rate = 0.000313810509397
Test accuracy 0.659375 Train accuracy 0.853125: Loss = 0.25782
Save steps: Test Accuracy 0.859375 is not higher than 0.921875
training: 100%|███████████████████████████| 3000/3000 [23:40<00:00, 1.27it/s]
    I tensorflow/core/common_runtime/gpu/gpu_device.cc:975] Creating
TensorFlow device (/gpu:0) -> (device: 0, name: GeForce GTX TITAN X, pci
bus id: 0000:01:00.0)
    ('Exported model at', '/home/ubuntu/models/pet-model/1')
```

至此我们拥有了一个测试准确率约为 92.18%的模型。另外，我们还将该模型导出为一个.pb 文件。此时，export_dir 文件夹具有以下结构。

```
- /home/ubuntu/models/
-- pet_model
---- 1
------ saved_model.pb
------ variables
```

9.5 在生产中利用模型提供服务

在生产中，我们需要创建一个端点，以便用户能够发送图片并接收结果。在 TensorFlow 中，可以通过 TensorFlow Serving 来我们的模型以提供服务。在本节中，我们将安装

TensorFlow Serving，并创建一个 Flask 应用程序，它允许用户通过一个 Web 界面上传图片。

9.5.1　设置 TensorFlow Serving

在生产服务器中，我们需要安装 TensorFlow Serving 并满足其使用条件。可以访问 TensorFlow Serving 的官方网站获取安装文件。接下来，我们将使用 TensorFlow Serving 提供的标准的 TensorFlow 模型服务器来为模型提供服务。首先，需要使用以下命令构建 tensorflow_model_server。

```
bazel build
//tensorflow_serving/model_servers:tensorflow_model_server
```

将训练服务器上路径/home/ubuntu/models/pet_model 中的所有文件复制到生产服务器中。将/home/ubuntu/productions 设置为存储所有生产模型的文件夹。productions 文件夹具有以下结构。

```
- /home/ubuntu/productions/
-- 1
---- saved_model.pb
---- variables
```

我们将使用 tmux 来保持模型服务器的运行。使用以下命令来安装 tmux。

```
sudo apt-get install tmux
```

使用以下命令运行一个 tmux 会话。

```
tmux new -s serving
```

在 tmux 会话中，将目录切换为 tensorflow_serving，并运行下面的命令。

```
bazel-bin/tensorflow_serving/model_servers/tensorflow_model_server --port=9000 --model_name=pet-model --model_base_path=/home/ubuntu/productions
```

控制台的输出如下所示。

```
    2017-05-29 13:44:32.203153: I
external/org_tensorflow/tensorflow/cc/saved_model/loader.cc:274] Loading
SavedModel: success. Took 537318 microseconds.
    2017-05-29 13:44:32.203243: I
tensorflow_serving/core/loader_harness.cc:86] Successfully loaded servable
version {name: pet-model version: 1}
    2017-05-29 13:44:32.205543: I
tensorflow_serving/model_servers/main.cc:298] Running ModelServer at
0.0.0.0:9000 ...
```

如你所见，该模型运行在主机 0.0.0.0 的端口 9000 上。在下一节中，我们将创建一个简单的 Python 客户端，并通过 gRPC 将一张图片发送给该服务器。

还应该注意到，当前的服务仅仅使用了生产服务器的 CPU。构建利用 GPU 的 TensorFlow Serving 超出了本章的范围。如果你想利用 GPU 提供服务，那么可能需要阅读第 12 章，第 12 章介绍了如何构建支持 GPU 的 TensorFlow 和 TensorFlow Serving。

9.5.2　运行和测试模型

项目代码库提供了一个名为 production 的包。在该包中，我们需要将 labels.txt 文件复制到数据集中，并创建一个新的 Python 文件 client.py，然后添加以下代码。

```python
import tensorflow as tf
import numpy as np
from tensorflow_serving.apis import prediction_service_pb2, predict_pb2
from grpc.beta import implementations
from scipy.misc import imread
from datetime import datetime


class Output:
    def __init__(self, score, label):
        self.score = score
        self.label = label

    def __repr__(self):
        return "Label: %s Score: %.2f" % (self.label, self.score)


def softmax(x):
    return np.exp(x) / np.sum(np.exp(x), axis=0)


def process_image(path, label_data, top_k=3):
    start_time = datetime.now()
    img = imread(path)

    host, port = "0.0.0.0:9000".split(":")
    channel = implementations.insecure_channel(host, int(port))
    stub = prediction_service_pb2.beta_create_PredictionService_stub(channel)
```

```python
request = predict_pb2.PredictRequest()
request.model_spec.name = "pet-model"
request.model_spec.signature_name = "predict_images"

request.inputs["images"].CopyFrom(
    tf.contrib.util.make_tensor_proto(
        img.astype(dtype=float),
        shape=img.shape, dtype=tf.float32
    )
)

result = stub.Predict(request, 20.)
scores = tf.contrib.util.make_ndarray(result.outputs["scores"])[0]
probs = softmax(scores)
index = sorted(range(len(probs)), key=lambda x: probs[x], reverse=True)

outputs = []
for i in range(top_k):
    outputs.append(Output(score=float(probs[index[i]]),
    label=label_data[index[i]]))

print(outputs)
print("total time", (datetime.now() -
start_time).total_seconds())
return outputs

if __name__ == "__main__":
label_data = [line.strip() for line in
open("production/labels.txt", 'r')]
process_image("samples_data/dog.jpg", label_data)
process_image("samples_data/cat.jpg", label_data)
```

在这段代码中，我们创建了一个 process_image 方法，它会从一个图像路径中读取图像，使用一些 TensorFlow 方法创建一个张量，并通过 gRPC 将其发送到模型服务器。我们还创建了一个 Output 类，这样就可以很容易地将它返回给 caller 方法。该方法的末尾，输出了结果和总时间，这样就可以更容易地调试它。可以运行该 client.py 文件来查看 process_image 是否正常工作。

python production/client.py

输出应该如下所示。

```
[Label: saint_bernard Score: 0.78, Label: american_bulldog Score: 0.21,
Label: staffordshire_bull_terrier Score: 0.00]
('total time', 14.943942)
[Label: Maine_Coon Score: 1.00, Label: Ragdoll Score: 0.00, Label:
Bengal Score: 0.00]
('total time', 14.918235)
```

结果正确。然而，每张图片的处理时间几乎长达 15s，这是因为我们是在 CPU 模式下使用的 TensorFlow Serving。正如前面所提到的，你可以在第 12 章中学习如何构建支持 GPU 的 TensorFlow Serving。如果按照教程进行操作，那么你将得到以下结果。

```
[Label: saint_bernard Score: 0.78, Label: american_bulldog Score: 0.21,
Label: staffordshire_bull_terrier Score: 0.00]
('total time', 0.493618)
[Label: Maine_Coon Score: 1.00, Label: Ragdoll Score: 0.00, Label:
Bengal Score: 0.00]
('total time', 0.023753)
```

第一次调用的处理时间约是 493 ms。但是，之后的调用时间只有约 23 ms，这比 CPU 版本快了很多。

9.5.3 设计 Web 服务器

在本节中，我们将创建一个 Flask 服务器，它允许用户上传图片，并且当模型给出错误的结果时，用户还可以设置正确的标签。production 包提供了所需的代码。然而，实现一个支持数据库的 Flask 服务器超出了本章的范围。在本节中，我们将描述 Flask 的所有要点，你可以更好地理解它。

表 9-4 描述了用户上传并纠正标签的主要流程。该流程是通过表 9-4 中的路由来实现的。

表 9-4

路 由	方 法	描 述
/	GET	该路由返回一个 Web 表单，供用户上传图片
/upload_image	POST	该路由从 POST 数据中获取图片，将其保存到上传目录，并调用 client.py 中的 process_image 函数来识别图像，然后保存结果到数据库
/results<result_id>	GET	该路由返回数据库中相应行的结果
/results<result_id>	POST	该路由将用户提供的标签保存到数据库，用于后续微调模型

续表

路　由	方　法	描　述
/user-labels	GET	该路由返回所有用户标记的图像列表。在微调过程中，我们将调用该路由以获取标记图像的列表
/model	POST	该路由允许训练服务器上的微调过程提供一个新的训练好的模型。该路由接收一个压缩模型链接、一个版本号、一个检查点名称和一个模型名称
/model	GET	该路由返回数据库中的最新模型。微调过程会调用该路由来知道哪个是最新模型，并对其进行微调

使用下面的命令在一个 tmux 会话中运行该服务器。

```
tmux new -s "flask"
python production/server.py
```

测试系统

现在，你可以通过 http://0.0.0.0:5000 访问服务器。此时，你将看到一个用于选择和提交图片的表单。

网站会将对应的图片和结果重定向到/results 页面，而此时用户的标签字段是空的。最后还有一个简短的表单，这样用户就可以为该模型提交正确的标签了。

9.6　在生产中自动化微调

系统运行一段时间后，将得到一些用户标记的图片。我们将创建一个微调进程，以实现系统每天自动运行，并利用新数据对最新的模型进行微调。

在 scripts 文件夹中创建一个名为 finetune.py 的文件，以保存微调进程的实现代码。

9.6.1　加载用户标记的数据

从生产服务器下载所有用户标记的图片的实现代码如下所示。

```python
import tensorflow as tf
import os
import json
import random
import requests
import shutil
```

```python
from scipy.misc import imread, imsave
from datetime import datetime
from tqdm import tqdm

import nets, models, datasets


def ensure_folder_exists(folder_path):
    if not os.path.exists(folder_path):
        os.mkdir(folder_path)
    return folder_path


def download_user_data(url, user_dir, train_ratio=0.8):
    response = requests.get("%s/user-labels" % url)
    data = json.loads(response.text)

    if not os.path.exists(user_dir):
        os.mkdir(user_dir)
    user_dir = ensure_folder_exists(user_dir)
    train_folder = ensure_folder_exists(os.path.join(user_dir, "trainval"))
    test_folder = ensure_folder_exists(os.path.join(user_dir, "test"))

    train_file = open(os.path.join(user_dir, 'trainval.txt'), 'w')
    test_file = open(os.path.join(user_dir, 'test.txt'), 'w')

    for image in data:
        is_train = random.random() < train_ratio
        image_url = image["url"]
        file_name = image_url.split("/")[-1]
        label = image["label"]
        name = image["name"]

        if is_train:
          target_folder = ensure_folder_exists(os.path.join(train_folder, name))
        else:
          target_folder = ensure_folder_exists(os.path.join(test_folder, name))

        target_file = os.path.join(target_folder, file_name) +
```

```python
            ".jpg"

        if not os.path.exists(target_file):
            response = requests.get("%s%s" % (url, image_url))
            temp_file_path = "/tmp/%s" % file_name
            with open(temp_file_path, 'wb') as f:
                for chunk in response:
                    f.write(chunk)

            image = imread(temp_file_path)
            imsave(target_file, image)
            os.remove(temp_file_path)
            print("Save file: %s" % target_file)

        label_path = "%s %s\n" % (label, target_file)
        if is_train:
            train_file.write(label_path)
        else:
            test_file.write(label_path)
```

在 `download_user_data` 中，调用/user-labels 端点来获取用户标记图片的列表。JSON 具有以下格式。

```
[
 {
  "id": 1,
  "label": 0,
  "name": "Abyssinian",
  "url": "/uploads/2017-05-23_14-56-45_Abyssinian-cat.jpeg"
 },
 {
  "id": 2,
  "label": 32,
  "name": "Siamese",
  "url": "/uploads/2017-05-23_14-57-33_fat-Siamese-cat.jpeg"
 }
]
```

在该 JSON 中，`label` 是用户选择的标签，`url` 是下载图片的链接。对于每一张图片，系统会将其下载到 tmp 文件夹中，并使用 scipy 中的 `imread` 和 `imsave` 方法保证图片为 JPEG 格式。与训练数据集相同，我们还会创建一个 trainval.txt 和 test.txt 文件。

9.6.2 对模型进行微调

为了对模型进行微调,我们需要知道哪个是最新的模型,以及恢复权重和偏差时对应的检查点。调用/model 端点来获取检查点名称和版本号。

```
def get_latest_model(url):
response = requests.get("%s/model" % url)
data = json.loads(response.text)
print(data)
return data["ckpt_name"], int(data["version"])
```

相应的 JSON 应该如下所示。

```
{
 "ckpt_name": "2017-05-26_02-12-49",
 "id": 10,
 "link": "http://1.53.110.161:8181/pet-model/8.zip",
 "name": "pet-model",
 "version": 8
}
```

现在,执行对模型微调的代码。首先从一些参数开始。

```
# Server info
URL = "http://localhost:5000"
dest_api = URL + "/model"

# Server Endpoints
source_api = "http://1.53.110.161:8181"

# Dataset
dataset_dir = "data/train_data"
user_dir = "data/user_data"
batch_size = 64
image_size = 224

# Learning rate
initial_learning_rate = 0.0001
decay_steps = 250
decay_rate = 0.9

# Validation
output_steps = 10  # Number of steps to print output
eval_steps = 20   # Number of steps to perform evaluations
```

```python
# Training
max_steps = 3000  # Number of steps to perform training
save_steps = 200  # Number of steps to perform saving
checkpoints
num_tests = 5  # Number of times to test for test accuracy
max_checkpoints_to_keep = 1
save_dir = "data/checkpoints"
train_vars = 'models/fc8-pets/weights:0,models/fc8-pets/biases:0'

# Get the latest model
last_checkpoint_name, last_version = get_latest_model(URL)
last_checkpoint_dir = os.path.join(save_dir, last_checkpoint_name)

# Export
export_dir = "/home/ubuntu/models/"
export_name = "pet-model"
export_version = last_version + 1
```

然后，执行微调循环。在以下代码中，我们将调用 `download_user_data` 下载所有用户标记的图片，并将 `user_dir` 传递给 `input_pipeline`，这样它就能加载新的图片。

```python
# Download user-labels data
download_user_data(URL, user_dir)

images, labels = datasets.input_pipeline(dataset_dir,
batch_size, is_training=True, user_dir=user_dir)
test_images, test_labels =
datasets.input_pipeline(dataset_dir, batch_size,
is_training=False, user_dir=user_dir)

 with tf.variable_scope("models") as scope:
 logits = nets.inference(images, is_training=True)
 scope.reuse_variables()
 test_logits = nets.inference(test_images, is_training=False)

total_loss = models.compute_loss(logits, labels)
train_accuracy = models.compute_accuracy(logits, labels)
test_accuracy = models.compute_accuracy(test_logits, test_labels)

global_step = tf.Variable(0, trainable=False)
```

```
learning_rate = models.get_learning_rate(global_step,
initial_learning_rate, decay_steps, decay_rate)
train_op = models.train(total_loss, learning_rate,
global_step, train_vars)

saver = tf.train.Saver(max_to_keep=max_checkpoints_to_keep)
checkpoint_name = datetime.now().strftime("%Y-%m-%d_%H-%M-%S")
checkpoints_dir = os.path.join(save_dir, checkpoint_name)
if not os.path.exists(save_dir):
  os.mkdir(save_dir)
if not os.path.exists(checkpoints_dir):
  os.mkdir(checkpoints_dir)

with tf.Session() as sess:
  sess.run(tf.global_variables_initializer())
  coords = tf.train.Coordinator()
  threads = tf.train.start_queue_runners(sess=sess,
  coord=coords)

saver.restore(sess,
models.get_model_path_from_ckpt(last_checkpoint_dir))
sess.run(global_step.assign(0))

last_saved_test_accuracy = 0
for i in range(num_tests):
    last_saved_test_accuracy += sess.run(test_accuracy)
last_saved_test_accuracy /= num_tests
should_export = False
print("Last model test accuracy
{}".format(last_saved_test_accuracy))
for i in tqdm(range(max_steps), desc="training"):
    _, loss_value, lr_value = sess.run([train_op, total_loss,
    learning_rate])

 if (i + 1) % output_steps == 0:
   print("Steps {}: Loss = {:.5f} Learning Rate =
   {}".format(i + 1, loss_value, lr_value))

    if (i + 1) % eval_steps == 0:
      test_acc, train_acc, loss_value =
      sess.run([test_accuracy, train_accuracy, total_loss])
        print("Test accuracy {} Train accuracy {} : Loss =
        {:.5f}".format(test_acc, train_acc, loss_value))
```

```python
        if (i + 1) % save_steps == 0 or i == max_steps - 1:
          test_acc = 0
          for i in range(num_tests):
            test_acc += sess.run(test_accuracy)
            test_acc /= num_tests

      if test_acc > last_saved_test_accuracy:
        print("Save steps: Test Accuracy {} is higher than
        {}".format(test_acc, last_saved_test_accuracy))
        last_saved_test_accuracy = test_acc
        saved_file = saver.save(sess,
      os.path.join(checkpoints_dir, 'model.ckpt'),
                                       global_step=global_step)
            should_export = True
            print("Save steps: Save to file %s " % saved_file)
          else:
            print("Save steps: Test Accuracy {} is not higher
      than {}".format(test_acc, last_saved_test_accuracy))

  if should_export:
    print("Export model with accuracy ",
    last_saved_test_accuracy)
    models.export_model(checkpoints_dir, export_dir,
    export_name, export_version)
    archive_and_send_file(source_api, dest_api,
    checkpoint_name, export_dir, export_name, export_version)
  coords.request_stop()
  coords.join(threads)
```

其他部分与训练循环非常相似。然而，我们并非从 caffe 模型中加载权重，而是使用最新模型的检查点，并多次运行测试以得到其测试准确率。

在微调循环的最后，需要一个新方法 archive_and_send_file 从导出的模型中生成一个归档文件，并将该链接发送给生产服务器。

```python
def make_archive(dir_path):
return shutil.make_archive(dir_path, 'zip', dir_path)


def archive_and_send_file(source_api, dest_api, ckpt_name,
export_dir, export_name, export_version):
model_dir = os.path.join(export_dir, export_name,
str(export_version))
```

```
    file_path = make_archive(model_dir)
    print("Zip model: ", file_path)

    data = {
        "link": "{}/{}/{}".format(source_api, export_name,
 str(export_version) + ".zip"),
        "ckpt_name": ckpt_name,
        "version": export_version,
        "name": export_name,
    }
    r = requests.post(dest_api, data=data)
    print("send_file", r.text)
```

注意，我们利用 source_api 参数创建了一个链接，这是到训练服务器 http://1.53.110.161:8181 的链接。我们将创建一个简单的 Apache 服务器来支持该功能。然而，在现实中，建议你将归档模型上传到云存储（如 Amazon S3）中。现在，我们将利用 Apache 展示最简单的方式。

使用以下命令来安装 Apache。

```
sudo apt-get install apache2
```

现在，在/etc/apache2/ports.conf 中的第 6 行，添加以下代码让 apache2 监听端口 8181。

```
Listen 8181
```

然后，在/etc/apache2/sites-available/000-default.conf 的开头添加以下代码以从/home/ubuntu/models 目录进行下载。

```
<VirtualHost *:8181>
  DocumentRoot "/home/ubuntu/models"
  <Directory />
    Require all granted
  </Directory>
</VirtualHost>
```

最后，重启 apache2 服务器。

```
sudo service apache2 restart
```

到目前为止，我们已经创建了微调所需要的所有代码。在第一次运行微调之前，需要向/model 端点发送一个 POST 请求，该请求包含了第一个模型的信息，这是因为我们已经将模型复制到了生产服务器上。

在 project 代码库中运行 finetune 脚本。

```
python scripts/finetune.py
```

控制台中的最后几行如下所示。

```
    Save steps: Test Accuracy 0.84 is higher than 0.916875
    Save steps: Save to file
data/checkpoints/2017-05-29_18-46-43/model.ckpt-2000
    ('Export model with accuracy ', 0.916875000000004)
    2017-05-29 18:47:31.642729: I
tensorflow/core/common_runtime/gpu/gpu_device.cc:977] Creating TensorFlow
device (/gpu:0) -> (device: 0, name: GeForce GTX TITAN X, pci bus id:
0000:01:00.0)
    ('Exported model at', '/home/ubuntu/models/pet-model/2')
    ('Zip model: ', '/home/ubuntu/models/pet-model/2.zip')
    ('send_file', u'{\n "ckpt_name": "2017-05-29_18-46-43", \n "id": 2,
\n  "link": "http://1.53.110.161:8181/pet-model/2.zip", \n "name": "pet-
model", \n  "version": 2\n}\n')
```

如你所见，新模型的测试准确率高于 91%。另外，该模型也导出并归档到/home/ubuntu/models/pet-model/2.zip 中。代码还调用了 /model 端点，以将链接发布到生产服务器。在生产服务器的 Flask 应用程序日志中，我们将得到以下结果。

```
('Start downloading', u'http://1.53.110.161:8181/pet-model/2.zip')
('Downloaded file at', u'/tmp/2.zip')
('Extracted at', u'/home/ubuntu/productions/2')
127.0.0.1 - - [29/May/2017 18:49:05] "POST /model HTTP/1.1" 200 -
```

这意味着 Flask 程序已经从训练服务器上下载了 2.zip 文件，并将文件内容提取到/home/ubuntu/productions/2 中。在 TensorFlow Serving 的 tmux 会话中，我们将得到以下结果。

```
    2017-05-29 18:49:06.234808: I
tensorflow_serving/core/loader_harness.cc:86] Successfully loaded servable
version {name: pet-model version: 2}
    2017-05-29 18:49:06.234840: I
tensorflow_serving/core/loader_harness.cc:137] Quiescing servable version
{name: pet-model version: 1}
    2017-05-29 18:49:06.234848: I
tensorflow_serving/core/loader_harness.cc:144] Done quiescing servable
version {name: pet-model version: 1}
    2017-05-29 18:49:06.234853: I
tensorflow_serving/core/loader_harness.cc:119] Unloading servable version
{name: pet-model version: 1}
    2017-05-29 18:49:06.240118: I
./tensorflow_serving/core/simple_loader.h:226] Calling
```

```
MallocExtension_ReleaseToSystem() with 645327546
    2017-05-29 18:49:06.240155: I
tensorflow_serving/core/loader_harness.cc:127] Done unloading servable
version {name: pet-model version: 1}
```

该输出表明，TensorFlow 模型服务器已经成功地加载了第 2 个版本的模型 pet-model，并卸载了版本 1。这也意味着我们已经提供了新模型，该模型在训练服务器上进行训练，并通过/model 端点发送到生产服务器上。

9.6.3 创建每天运行的 cronjob

最后，需要将微调程序设置成每天运行，并可以自动将新模型上传到服务器上。在训练服务器上创建一个 `crontab`，就可以轻松实现这一目的。

首先，需要运行 `crontab` 命令。

crontab -e

然后，添加以下行来定义想要 finetune.py 运行的时间点。

0 3 * * * python /home/ubuntu/project/scripts/finetune.py

按照我们的定义，Python 命令将会在每天凌晨 3 点整运行。

9.7 总结

在本章中，通过训练和提供深度学习模型，我们实现了一个完整的、真实的生产模型。我们还在 Flask 应用程序中创建了一个 Web 界面，以允许用户上传他们的图片并接收结果。此外，我们的模型可以实现每天自动调优以提高系统的质量。不过，可以考虑通过以下几点来改善整个系统。

- 模型和检查点应该保存在云存储中。
- Flask 应用和 TensorFlow Serving 应该由另一个更好的进程管理系统来管理，例如 Supervisor。
- 应该有一个 Web 界面，以便团队能够批准用户选择的标签。我们不应该完全依靠用户来决定训练集。
- 应该创建支持 GPU 的 TensorFlow Serving，以获取最佳的性能。

第 10 章 系统上线

在本章，我们将学习更多关于亚马逊 Web 服务（Amazon Web Services，AWS）的知识，以及如何创建一个深度神经网络来解决视频动作识别问题。此外，本章还将介绍如何使用多个 GPU 加快训练过程。在本章最后，我们将快速浏览亚马逊 Mechanical Turk 服务（Amazon Mechanical Turk Service），通过它我们可以收集标签并纠正模型结果。

10.1 快速浏览亚马逊 Web 服务

亚马逊 Web 服务是最受欢迎的云平台之一，由亚马逊（Amazon）公司开发。它提供了很多服务，如云计算、存储、数据库服务、内容分发以及其他服务。在本节中，我们仅关注 Amazon EC2 提供的虚拟服务器服务。Amazon EC2 允许用户创建多个服务器，这些服务器支持模型的运行和训练流程。当 Amazon EC2 服务于终端用户模型时，用户可以阅读第 9 章来学习 TensorFlow Serving。在训练流程方面，Amazon EC2 提供了很多可以使用的实例类型。我们可以使用 Amazon EC2 的 CPU 服务器运行 Web 机器人以从互联网上收集数据。另外，Amazon EC2 也存在几种拥有多个 NVIDIA GPU 的实例类型。

Amazon EC2 提供了各种各样的实例类型来适用不同的使用场景。这些实例类型分为 5 种类别，如下所示。

- ◆ 一般用途。
- ◆ 计算优化。
- ◆ 内存优化。
- ◆ 存储优化。

◆ 加速计算实例。

前 4 种类别适合运行后端服务器。加速计算实例拥有多个 NVIDIA GPU，可用于服务模型或者使用高端的 GPU 训练新模型。加速计算实例存在 3 种类型——P2、G2 和 F1。

10.1.1 P2 实例

P2 实例包含高性能的 NVIDIA K80 GPUs，每个 GPU 都拥有 2 496 个 CUDA 核心和 12 GB 的 GPU 内存。P2 实例有 3 种型号，如表 10-1 所示。

表 10-1

型号	GPU 数量	vCPU	CPU 内存（GB）	GPU 内存（GB）
p2.xlarge	1	4	61	12
p2.8xlarge	8	32	488	96
p2.16xlarge	16	64	732	192

这些具有大 GPU 内存的实例适用于训练模型。有了更多的 GPU 内存，我们可以使用更大的批量尺寸和包含大量参数的神经网络来训练模型。

10.1.2 G2 实例

G2 实例包含高性能的 NVIDIA GPUs，每个都拥有 1 536 个 CUDA 核心和 4 GB 的 GPU 内存。G2 实例有两种型号，如表 10-2 所示。

表 10-2

型号	GPU 数量	vCPU	CPU 内存（GB）	SSD 存储（GB）
g2.2xlarge	1	8	15	1 × 60
g2.8xlarge	4	32	60	2 × 120

这些型号的实例的 GPU 内存为 4 GB，所以它们在训练中会有所限制。然而，4 GB 的 GPU 内存通常足以为终端用户模型提供服务。选择 G2 实例的一个最重要的因素是，G2 实例比 P2 实例更便宜，这就使得用户能够在一个负载均衡器下部署多个服务器，以实现较高的可扩展性。

10.1.3 F1 实例

F1 实例支持现场可编程门阵列（Field Programmable Gate Array，FPGA）。F1 实例存在两种型号，如表 10-3 所示。

表 10-3

型 号	GPU 数量	vCPU	CPU 内存（GB）	SSD 存储（GB）
f1.2xlarge	1	8	122	470
f1.16xlarge	8	64	976	4 × 940

具有大内存和强计算能力的 FPGA 在深度学习领域很有前景。然而，TensorFlow 和其他流行的深度学习库目前并不支持 FPGA。因此，在下一节中，我们只讨论 P2 和 G2 实例的价格。

10.1.4　定价

Amazon EC2 为实例提供了 3 个定价选项——按需实例（on-demand instance）、保留实例（reserved instance）和竞价实例（spot instance）。

◆ 按需实例可以让你在不中断的情况下运行服务器。如果你只想使用几天或几周，那么这种实例比较合适。

◆ 保留实例为你提供了保留的选项，可以将该实例保留 1 年或 3 年。与按需实例相比，这种实例有很大的折扣。如果你想将服务器用于生产环境，那么这种实例比较合适。

◆ 竞价实例提供了为服务器竞价的选项。你可以选择愿意每小时为每个实例所支付的最高价格。这种实例可以为你节省很多钱。然而，如果其他人出价更高，那么这些实例随时都有可能会被终止。如果你的系统能够处理中断，或者你只是想探索研究某种服务，那么这种实例就比较合适。

用户可以单击 **Add New Row** 按钮并选择一种实例类型。

在图 10-1 中，我们选择了一个 p2.xlarge 服务器。在本书编写之际，其一个月的价格是 658.80 美元。

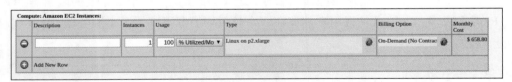

图 10-1

单击 **Billing Option** 列，你将看到一个 p2.xlarge 服务器保留实例的价格，如图 10-2 所示。

Select	Name	Upfront Price	Effective Hourly Cost	Effective Monthly Cost	1 Year Cost	3 Year Cost
●	On-Demand (No Contract)	- - -	0.900	658.80	7905.60	23716.80
○	1 Yr No Upfront Reserved	0.00	0.614	448.22	5378.64	16135.92
○	1 Yr Partial Upfront Reserved	2562.00	0.585	427.02	5124.24	15372.72
○	1 Yr All Upfront Reserved	5022.00	0.573	418.50	5022.00	15066.00
○	3 Yr Partial Upfront Reserved	5584.00	0.425	310.24	- - -	11168.32
○	3 Yr All Upfront Reserved	10499.00	0.399	291.64	- - -	10499.00
○	3 Yr No Upfront Convertible	0.00	0.528	385.44	- - -	13875.84
○	3 Yr Partial Upfront Convertible	6422.00	0.488	356.51	- - -	12834.32
○	3 Yr All Upfront Convertible	12588.00	0.479	349.67	- - -	12588.00

图 10-2

此外，Amazon EC2 还有很多其他实例类型。建议看一下其他类型的实例，然后选择最符合你需求的服务器。

在下一节中，我们将创建一个可以利用 TensorFlow 执行视频动作识别的新模型。另外，我们还将使用多个 GPU 来提高训练性能。

10.2 应用程序概述

在计算机视觉和机器学习领域，人体动作识别是一个非常有趣的问题。针对该问题有两种常见的方法：静态图像动作识别和视频动作识别。在静态图像动作识别中，我们可以微调一个来自 ImageNet 的预训练模型，并基于静态图像对动作进行分类。读者可以查看前面的章节以获取更多信息。在本章中，我们将创建一个可以从视频中识别人体动作的模型。在本章结尾，我们将展示如何使用多个 GPU 来加速训练过程。

10.2.1 数据集

在训练过程中，用户可以使用很多可用的数据集，如下所示。

- ◆ UCF101 是一个动作识别数据集，它包含 101 个动作类别的真实动作视频，共计 13 320 段视频。这使得该数据集成为很多论文研究很好的选择。
- ◆ ActivityNet 是一个用于描述人体活动的大型数据集。其中包含 200 个类别、超过 648 小时的视频，每个类别大约有 100 个视频。
- ◆ Sports-1M 是另一个用于运动识别的大型数据集。其中总共有 1 133 158 个视频，并标注了 487 个运动标签。

在本章中，我们将使用 UCF101 来执行训练过程。此外，推荐读者尝试将本章中讨论的技术应用到一个大型数据集上，以充分利用多 GPU 训练。

10.2.2 准备数据集和输入管道

UCF101 数据集包含 101 个动作类别，例如篮球投射、弹吉他以及冲浪等。你需要先下载 UCF101 数据集 UCF101.rar，以及用于动作识别的训练/测试子集 UCF101TrainTestSplits-RecognitionTask.zip。在进入下一节之前，你需要提取这些数据集。在下一节中，我们将在训练之前对视频进行预处理。

1. 预处理训练视频

UCF101 包含 13 320 个帧率和分辨率分别固定为 25 帧/秒和 320 × 240 的剪辑视频。所有的剪辑视频都以 AVI 格式存储，在 TensorFlow 中使用它们并不方便。因此，在本节中，我们将从所有视频中提取视频帧并将其保存为 JPEG 文件。我们只会以 4 帧/秒的固定帧率来提取视频帧，这样可以减小网络的输入大小。

在开始实现代码之前，需要安装 PyAV 库。

首先，在 root 文件夹中创建一个名为 scripts 的 Python 包。然后，在 scripts 目录中创建一个新的 Python 文件 convert_ucf101.py。在该文件中添加第一批代码来导入和定义一些参数，如下所示。

```
import av
import os
import random
import tensorflow as tf
from tqdm import tqdm

FLAGS = tf.app.flags.FLAGS
tf.app.flags.DEFINE_string(
    'dataset_dir', '/mnt/DATA02/Dataset/UCF101',
    'The folder that contains the extracted content of UCF101.rar'
)

tf.app.flags.DEFINE_string(
    'train_test_list_dir',
'/mnt/DATA02/Dataset/UCF101/ucfTrainTestlist',
    'The folder that contains the extracted content of
UCF101TrainTestSplits-RecognitionTask.zip'
)
```

```python
tf.app.flags.DEFINE_string(
    'target_dir', '/home/ubuntu/datasets/ucf101',
    'The location where all the images will be stored'
)

tf.app.flags.DEFINE_integer(
    'fps', 4,
    'Framerate to export'
)

def ensure_folder_exists(folder_path):
    if not os.path.exists(folder_path):
        os.mkdir(folder_path)

    return folder_path
```

在上述代码中，dataset_dir 和 train_test_list_dir 分别表示从 UCF101.rar 和 UCF101TrainTestSplits-RecognitionTask.zip 中提取的内容所存放的文件夹的位置。target_dir 是存放所有训练图片的文件夹。ensure_folder_exists 是一个工具函数。如果某个文件夹不存在，那么该函数就会创建该文件夹。

接下来，定义 Python 代码的 main 函数。

```python
def main(_):
    if not FLAGS.dataset_dir:
        raise ValueError("You must supply the dataset directory with
 --dataset_dir")

    ensure_folder_exists(FLAGS.target_dir)
    convert_data(["trainlist01.txt", "trainlist02.txt", "trainlist03.txt"], training=True)
    convert_data(["testlist01.txt", "testlist02.txt", "testlist03.txt"], training=False)

if __name__ == "__main__":
    tf.app.run()
```

在 main 函数中，我们创建了 target_dir 文件夹，并调用了 convert_data 函数（稍后就会创建该函数）。convert_data 函数用于获取数据集中的训练/测试文本文件列表，以及一个名为 training 的布尔变量。这个布尔变量用于表示文本文件是否用于训练过程。

下面是其中一个文本文件中的部分内容。

ApplyEyeMakeup/v_ApplyEyeMakeup_g08_c01.avi 1
ApplyEyeMakeup/v_ApplyEyeMakeup_g08_c02.avi 1
ApplyEyeMakeup/v_ApplyEyeMakeup_g08_c03.avi 1

文本文件的每一行都包含了视频文件路径和正确的标签。在该示例中，有 3 个 ApplyEyeMakeup 类别的视频路径，而该类别是数据集中的第 1 类。

此处的主要思路是，读取文本文件的每一行，以 JPEG 格式提取视频帧，并利用对应的标签来保存所提取文件的位置，以供进一步训练。下面是 `convert_data` 函数的实现代码。

```python
def convert_data(list_files, training=False):
    lines = []
    for txt in list_files:
        lines += [line.strip() for line in open(os.path.join(FLAGS.train_test_list_dir, txt))]

    output_name = "train" if training else "test"

    random.shuffle(lines)

    target_dir = ensure_folder_exists(os.path.join(FLAGS.target_dir, output_name))
    class_index_file = os.path.join(FLAGS.train_test_list_dir, "classInd.txt")
    class_index = {line.split(" ")[1].strip(): int(line.split(" ")[0]) - 1 for line in open(class_index_file)}

    with open(os.path.join(FLAGS.target_dir, output_name + ".txt"), "w") as f:
        for line in tqdm(lines):
            if training:
                filename, _ = line.strip().split(" ")
            else:
                filename = line.strip()
            class_folder, video_name = filename.split("/")

            label = class_index[class_folder]
            video_name = video_name.replace(".avi", "")
            target_class_folder =
```

10.2 应用程序概述

```
ensure_folder_exists(os.path.join(target_dir, class_folder))
            target_folder =
ensure_folder_exists(os.path.join(target_class_folder,
video_name))

            container = av.open(os.path.join(FLAGS.dataset_dir,
            filename))
            frame_to_skip = int(25.0 / FLAGS.fps)
            last_frame = -1
            frame_index = 0
            for frame in container.decode(video=0):
                if last_frame < 0 or frame.index > last_frame +
                frame_to_skip:
                    last_frame = frame.index
                    image = frame.to_image()
                    target_file = os.path.join(target_folder,
                "%04d.jpg" % frame_index)
                    image.save(target_file)
                    frame_index += 1
            f.write("{} {} {}\n".format("%s/%s" % (class_folder,
            video_name), label, frame_index))

    if training:
        with open(os.path.join(FLAGS.target_dir, "label.txt"), "w")
        as f:
            for class_name in sorted(class_index,
            key=class_index.get):
                f.write("%s\n" % class_name)
```

从上述代码可以看出，我们从文本文件中加载视频路径，并使用 PyAV 库打开 AVI 文件。然后，使用 `FLAGS.fps` 来控制每秒提取的帧数。可以使用下面的命令运行 scripts/convert_ucf101.py 文件。

python scripts/convert_ucf101.py

转换所有的视频大约需要 30 分钟。最后，`target_dir` 文件夹中将包含以下文件。

label.txt   test   test.txt   train   train.txt

`train.txt` 文件的内容如下所示。

Punch/v_Punch_g25_c03 70 43
Haircut/v_Haircut_g20_c01 33 36
BrushingTeeth/v_BrushingTeeth_g25_c02 19 33
Nunchucks/v_Nunchucks_g03_c04 55 36

```
BoxingSpeedBag/v_BoxingSpeedBag_g16_c04 17 21
```
这种格式可以按以下方式来理解。

```
<Folder location of the video> <Label> <Number of frames in the folder>
```

有一点必须记住,那就是 `train.txt` 和 `test.txt` 中的标签值范围为 0~100,而 UCF101 中的标签值范围是 1~101。这是因为在 TensorFlow 中,`sparse_softmax_cross_entropy` 函数要求类别标签从 0 开始。

2. 输入管道 RandomShuffleQueue

如果阅读过第 9 章,你就会知道可以在 TensorFlow 中使用 TextLineReader 来逐行读取文本文件,而且可以直接在 TensorFlow 中读取图片。然而,当数据中仅仅包含文件夹位置和标签时,事情就会变得更加复杂。此外,我们只需要某个文件夹中所有帧的一个子集。例如,如果帧的数量是 30,而我们只需要其中的 10 帧来进行训练,那么就需要获取一个 0~20 的随机值,并从其中选择 10 帧。在本章中,我们将使用另一种机制,以纯 Python 代码来对视频帧进行采样,并将选定的帧路径输入到 RandomShuffleQueue 中进行训练。此外,还会使用 tf.train.batch_join 和多个预处理线程来改善训练过程。

首先,在 root 文件夹中创建一个新 Python 文件 utils.py,并添加以下代码。

```python
def lines_from_file(filename, repeat=False):
    with open(filename) as handle:
        while True:
            try:
                line = next(handle)
                yield line.strip()
            except StopIteration as e:
                if repeat:
                    handle.seek(0)
                else:
                    raise

if __name__ == "__main__":
    data_reader = lines_from_file("/home/ubuntu/datasets/ucf101/train.txt", repeat=True)

    for i in range(15):
        print(next(data_reader))
```

这段代码创建了一个名为 `lines_from_file` 的生成器函数来逐行读取文本文件。这段代码还添加了一个 `repeat` 参数,以便在到达文件末尾时生成器函数可以从头开始读取文本。

我们已经添加了 main 部分，你可以尝试运行它以查看生成器是如何工作的。

python utils.py

现在，在 root 文件夹下创建一个新 Python 文件 datasets.py，并添加以下代码。

```python
import tensorflow as tf
import cv2
import os
import random

from tensorflow.python.ops import data_flow_ops
from utils import lines_from_file

def sample_videos(data_reader, root_folder, num_samples, num_frames):
    image_paths = list()
    labels = list()
    while True:
        if len(labels) >= num_samples:
            break
        line = next(data_reader)
        video_folder, label, max_frames = line.strip().split(" ")
        max_frames = int(max_frames)
        label = int(label)
        if max_frames > num_frames:
            start_index = random.randint(0, max_frames - num_frames)
            frame_paths = list()
            for index in range(start_index, start_index + num_frames):
                frame_path = os.path.join(root_folder, video_folder, "%04d.jpg" % index)
                frame_paths.append(frame_path)
            image_paths.append(frame_paths)
            labels.append(label)
    return image_paths, labels

if __name__ == "__main__":
    num_frames = 5
    root_folder = "/home/ubuntu/datasets/ucf101/train/"
    data_reader = lines_from_file("/home/ubuntu/datasets/ucf101/train.txt",
```

```
    repeat=True)
image_paths, labels = sample_videos(data_reader,
root_folder=root_folder,
num_samples=3,
num_frames=num_frames)
    print("image_paths", image_paths)
    print("labels", labels)
```

sample_videos 函数接收 lines_from_file 函数返回的生成器对象，并使用 next 函数获取所需的样本。可以看到，我们使用 random.randint 方法来随机化起始帧的位置。

可以使用下面的命令运行 main 部分，以查看 sample_videos 是如何工作的。

python datasets.py

到目前为止，我们已经将数据集文本文件读取到 image_paths 和 labels 变量中，它们都是 Python 列表。在稍后的训练流程中，我们将使用 TensorFlow 内置的 RandomShuffleQueue 把 image_paths 和 labels 存入到该队列。

现在，我们需要创建一个方法。该方法将会在训练过程中从 RandomShuffleQueue 中获取数据，并在多个线程中执行预处理，以及将数据发送到 batch_join 函数中以创建用于训练的小批量数据。

在 dataset.py 文件中，添加以下代码。

```
def input_pipeline(input_queue, batch_size=32, num_threads=8,
image_size=112):
    frames_and_labels = []
    for _ in range(num_threads):
        frame_paths, label = input_queue.dequeue()
        frames = []
        for filename in tf.unstack(frame_paths):
            file_contents = tf.read_file(filename)
            image = tf.image.decode_jpeg(file_contents)
            image = _aspect_preserving_resize(image, image_size)
            image = tf.image.resize_image_with_crop_or_pad(image,
            image_size, image_size)
            image = tf.image.per_image_standardization(image)
            image.set_shape((image_size, image_size, 3))
            frames.append(image)
        frames_and_labels.append([frames, label])

    frames_batch, labels_batch = tf.train.batch_join(
```

```
            frames_and_labels, batch_size=batch_size,
            capacity=4 * num_threads * batch_size,
        )
        return frames_batch, labels_batch
```

在这段代码中，我们准备了一个名为 frames_and_labels 的数组，并使用一个迭代 num_threads 次的 for 循环，这是向预处理过程添加多线程支持的一种非常方便的方式。在每个线程中，我们将调用 input_queue 的 dequeue 方法来获取 frame_paths 和 label。从上一节的 sample_video 函数，我们知道 frame_paths 是一个保存所选视频帧的列表。因此，我们使用另一个 for 循环来循环遍历每一帧。在每一帧中，我们读取、调整图像大小并执行图像标准化，这部分代码类似于第 9 章中的代码。在输入管道的末尾，我们添加了 frames_and_labels 和 batch_size 参数。返回的 frames_batch 和 labels_batch 将用于后面的训练。

最后，还应该添加以下代码，其中包含了 _aspect_preserving_resize 函数。

```
def _smallest_size_at_least(height, width, smallest_side):
    smallest_side = tf.convert_to_tensor(smallest_side, dtype=tf.int32)

    height = tf.to_float(height)
    width = tf.to_float(width)
    smallest_side = tf.to_float(smallest_side)

    scale = tf.cond(tf.greater(height, width),
                    lambda: smallest_side / width,
                    lambda: smallest_side / height)
    new_height = tf.to_int32(height * scale)
    new_width = tf.to_int32(width * scale)
    return new_height, new_width


def _aspect_preserving_resize(image, smallest_side):
    smallest_side = tf.convert_to_tensor(smallest_side, dtype=tf.int32)
    shape = tf.shape(image)
    height = shape[0]
    width = shape[1]
    new_height, new_width = _smallest_size_at_least(height, width, smallest_side)
    image = tf.expand_dims(image, 0)
    resized_image = tf.image.resize_bilinear(image, [new_height,
```

```
new_width], align_corners=False)
    resized_image = tf.squeeze(resized_image)
    resized_image.set_shape([None, None, 3])
    return resized_image
```

这段代码和在第 9 章中使用的代码是一样的。

在下一节，我们将创建深度神经网络架构，并使用它执行 101 个类别的视频动作识别。

10.2.3 神经网络架构

在本节中，我们将创建一个神经网络，该神经网络将接收 10 个视频帧并输出分别属于 101 个动作类别的概率。我们将基于 TensorFlow 中的 conv3d 操作创建一个神经网络。该神经网络的灵感来自 D.Tran 等人的论文"Learning Spatiotemporal Features with 3D Convolutional Networks"。然而，我们简化了该模型，以便更容易理解。此外，我们还采用了 D.Tran 等人未提及的一些技术，例如批量归一化和 dropout。

现在，创建一个新的 Python 文件 nets.py，并输入以下代码。

```
import tensorflow as tf
from utils import print_variables, print_layers
from tensorflow.contrib.layers.python.layers.layers import
batch_norm
def inference(input_data, is_training=False):
    conv1 = _conv3d(input_data, 3, 3, 3, 64, 1, 1, 1, "conv1")
    pool1 = _max_pool3d(conv1, 1, 2, 2, 1, 2, 2, "pool1")

    conv2 = _conv3d(pool1, 3, 3, 3, 128, 1, 1, 1, "conv2")
    pool2 = _max_pool3d(conv2, 2, 2, 2, 2, 2, 2, "pool2")
    conv3a = _conv3d(pool2, 3, 3, 3, 256, 1, 1, 1, "conv3a")
    conv3b = _conv3d(conv3a, 3, 3, 3, 256, 1, 1, 1, "conv3b")
    pool3 = _max_pool3d(conv3b, 2, 2, 2, 2, 2, 2, "pool3")
    conv4a = _conv3d(pool3, 3, 3, 3, 512, 1, 1, 1, "conv4a")
    conv4b = _conv3d(conv4a, 3, 3, 3, 512, 1, 1, 1, "conv4b")
    pool4 = _max_pool3d(conv4b, 2, 2, 2, 2, 2, 2, "pool4")
    conv5a = _conv3d(pool4, 3, 3, 3, 512, 1, 1, 1, "conv5a")
    conv5b = _conv3d(conv5a, 3, 3, 3, 512, 1, 1, 1, "conv5b")
    pool5 = _max_pool3d(conv5b, 2, 2, 2, 2, 2, 2, "pool5")

    fc6 = _fully_connected(pool5, 4096, name="fc6")
    fc7 = _fully_connected(fc6, 4096, name="fc7")

    if is_training:
```

```
        fc7 = tf.nn.dropout(fc7, keep_prob=0.5)
        fc8 = _fully_connected(fc7, 101, name='fc8', relu=False)
        endpoints = dict()
        endpoints["conv1"] = conv1
        endpoints["pool1"] = pool1
        endpoints["conv2"] = conv2
        endpoints["pool2"] = pool2
        endpoints["conv3a"] = conv3a
        endpoints["conv3b"] = conv3b
        endpoints["pool3"] = pool3
        endpoints["conv4a"] = conv4a
        endpoints["conv4b"] = conv4b
        endpoints["pool4"] = pool4
        endpoints["conv5a"] = conv5a
        endpoints["conv5b"] = conv5b
        endpoints["pool5"] = pool5
        endpoints["fc6"] = fc6
        endpoints["fc7"] = fc7
        endpoints["fc8"] = fc8
        return fc8, endpoints

if __name__ == "__main__":
    inputs = tf.placeholder(tf.float32, [None, 10, 112, 112, 3], name="inputs")
    outputs, endpoints = inference(inputs)

    print_variables(tf.global_variables())
    print_variables([inputs, outputs])
    print_layers(endpoints)
```

在 inference 函数中,我们调用_conv3d、_max_pool3d 和_fully_connected 来创建网络,这和前面章节中处理图像的 CNN 网络没有太大的不同。在函数的末尾,我们还创建了一个名为 endpoints 的字典。该字典将在 main 部分中用于可视化网络架构。

接下来,添加_conv3d 和_max_pool3d 函数的代码。

```
def _conv3d(input_data, k_d, k_h, k_w, c_o, s_d, s_h, s_w, name, relu=True, padding="SAME"):
    c_i = input_data.get_shape()[-1].value
    convolve = lambda i, k: tf.nn.conv3d(i, k, [1, s_d, s_h, s_w, 1], padding=padding)
    with tf.variable_scope(name) as scope:
        weights = tf.get_variable(name="weights", shape=[k_d, k_h, k_w, c_i, c_o],
```

```
regularizer = tf.contrib.layers.l2_regularizer(scale=0.0001),
initializer=tf.truncated_normal_initializer(stddev=1e-1,
dtype=tf.float32))
        conv = convolve(input_data, weights)
        biases = tf.get_variable(name="biases",
shape=[c_o], dtype=tf.float32,
initializer = tf.constant_initializer(value=0.0))
        output = tf.nn.bias_add(conv, biases)
        if relu:
            output = tf.nn.relu(output, name=scope.name)
        return batch_norm(output)


def _max_pool3d(input_data, k_d, k_h, k_w, s_d, s_h, s_w, name,
padding="SAME"):
    return tf.nn.max_pool3d(input_data,
ksize=[1, k_d, k_h, k_w, 1],
strides=[1, s_d, s_h, s_w, 1], padding=padding, name=name)
```

这段代码之前讲过。然而，此处使用的是内置的 `tf.nn.conv3d` 和 `tf.nn.max_pool3d` 函数，而不是用于图像的 `tf.nn.conv2d` 和 `tf.nn.max_pool3d` 函数。添加参数 `k_d` 和 `s_d` 来提供过滤器的深度信息。此外，还需要从零开始对该网络进行训练，而不是使用任何预训练模型。使用 `batch_norm` 函数向每层添加批量归一化操作。

下面为全连接层添加代码，具体如下。

```
def _fully_connected(input_data, num_output, name, relu=True):
    with tf.variable_scope(name) as scope:
        input_shape = input_data.get_shape()
        if input_shape.ndims == 5:
            dim = 1
            for d in input_shape[1:].as_list():
                dim *= d
            feed_in = tf.reshape(input_data, [-1, dim])
        else:
            feed_in, dim = (input_data, input_shape[-1].value)
        weights = tf.get_variable(name="weights",
shape=[dim, num_output],
regularizer = tf.contrib.layers.l2_regularizer(scale=0.0001),
initializer=tf.truncated_normal_initializer(stddev=1e-1,
dtype=tf.float32))
        biases = tf.get_variable(name="biases",
shape=[num_output], dtype=tf.float32,
```

```
        initializer=tf.constant_initializer(value=0.0))
    op = tf.nn.relu_layer if relu else tf.nn.xw_plus_b
    output = op(feed_in, weights, biases, name=scope.name)
    return batch_norm(output)
```

该函数与之前用于处理图像的方法有点不同。首先，检查 input_shape.ndims 的值，它应该等于 5 而不是 4。其次，为输出添加批量归一化操作。最后，打开 utils.py 文件，添加以下函数。

```
from prettytable import PrettyTable
def print_variables(variables):
    table = PrettyTable(["Variable Name", "Shape"])
    for var in variables:
        table.add_row([var.name, var.get_shape()])
    print(table)
    print("")

def print_layers(layers):
    table = PrettyTable(["Layer Name", "Shape"])
    for var in layers.values():
        table.add_row([var.name, var.get_shape()])
    print(table)
    print("")
```

现在，运行 nets.py 以更好地理解网络架构。

python nets.py

在控制台结果的第一部分中，你会看到表 10-4 所示的结果。

表 10-4

Variable Name	Shape
conv1/weights:0	(3, 3, 3, 3, 64)
conv1/biases:0	(64,)
conv1/BatchNorm/beta:0	(64,)
conv1/BatchNorm/moving_mean:0	(64,)
conv1/BatchNorm/moving_variance:0	(64,)
...	...
fc8/weights:0	(4096, 101)
fc8/biases:0	(101,)
fc8/BatchNorm/beta:0	(101,)
fc8/BatchNorm/moving_mean:0	(101,)
fc8/BatchNorm/moving_variance:0	(101,)

这些是网络中 `Variables` 的信息。如你所见，包含文本 `BatchNorm` 的 3 个变量都被添加到了每一层中。这些变量增加了网络需要学习的参数的总数量。然而，因为我们需要从头开始训练，所以在没有批量归一化的情况下，网络训练将会变得更加困难。另外，批处理归一化也增强了网络对未知数据进行调整的能力。

控制台的第二张表如表 10-5 所示。

表 10-5

Variable Name	Shape
inputs:0	(?, 10, 112, 112, 3)
fc8/BatchNorm/batchnorm/add_1:0	(?, 101)

表 10-5 所示的是网络输入和输出的尺寸。如你所见，输入包含 10 个大小为（112,112,3）的视频帧，输出包含一个元素数量为 101 的向量。

在表 10-6 中，你将会看到网络中每层的输出的尺寸是如何变化的。

表 10-6

Layer Name	Shape
fc6/BatchNorm/batchnorm/add_1:0	(?, 4096)
fc7/BatchNorm/batchnorm/add_1:0	(?, 4096)
fc8/BatchNorm/batchnorm/add_1:0	(?, 101)
...	...
conv1/BatchNorm/batchnorm/add_1:0	(?, 10, 112, 112, 64)
conv2/BatchNorm/batchnorm/add_1:0	(?, 10, 56, 56, 128)

在前文表格中可以看到，conv1 层的输出尺寸与输入尺寸相等，而由于最大池化的影响，conv2 层的输出发生了变化。

现在，创建一个新的 Python 文件 models.py，并添加以下代码。

```
import tensorflow as tf

def compute_loss(logits, labels):
    labels = tf.squeeze(tf.cast(labels, tf.int32))

    cross_entropy = tf.nn.sparse_softmax_cross_entropy_with_logits(logits=logits, labels=labels)
    cross_entropy_loss= tf.reduce_mean(cross_entropy)
```

```
    reg_loss = 
tf.reduce_mean(tf.get_collection(tf.GraphKeys.REGULARIZATION_LOSSES
))

    return cross_entropy_loss + reg_loss, cross_entropy_loss,
reg_loss


def compute_accuracy(logits, labels):
    labels = tf.squeeze(tf.cast(labels, tf.int32))
    batch_predictions = tf.cast(tf.argmax(logits, 1), tf.int32)
    predicted_correctly = tf.equal(batch_predictions, labels)
    accuracy = tf.reduce_mean(tf.cast(predicted_correctly,
tf.float32))
    return accuracy


def get_learning_rate(global_step, initial_value, decay_steps,
decay_rate):
    learning_rate = tf.train.exponential_decay(initial_value,
    global_step, decay_steps, decay_rate, staircase=True)
    return learning_rate


def train(total_loss, learning_rate, global_step):
    optimizer = tf.train.AdamOptimizer(learning_rate)
    train_op = optimizer.minimize(total_loss, global_step)
    return train_op
```

这些函数创建了计算损失值、准确率和学习率的相应操作，并执行了训练过程。上述函数和上一章中的是相同的，此处不再解释。

现在，我们有了训练网络识别视频动作所需要的所有函数。在下一节中，我们将在单个 GPU 上开始训练，并在 TensorBoard 中将结果可视化。

10.2.4 单 GPU 训练流程

在 scripts 包中，创建一个新的 Python 文件 train.py。

首先，定义参数。

```
import tensorflow as tf
import os
import sys
```

```python
from datetime import datetime
from tensorflow.python.ops import data_flow_ops

import nets
import models
from utils import lines_from_file
from datasets import sample_videos, input_pipeline

# Dataset
num_frames = 16
train_folder = "/home/ubuntu/datasets/ucf101/train/"
train_txt = "/home/ubuntu/datasets/ucf101/train.txt"

# Learning rate
initial_learning_rate = 0.001
decay_steps = 1000
decay_rate = 0.7
# Training
image_size = 112
batch_size = 24
num_epochs = 20
epoch_size = 28747

train_enqueue_steps = 100
min_queue_size = 1000

save_steps = 200  # Number of steps to perform saving checkpoints
test_steps = 20  # Number of times to test for test accuracy
start_test_step = 50

max_checkpoints_to_keep = 2
save_dir = "/home/ubuntu/checkpoints/ucf101"
```

这些参数都是非常直观的。现在，我们为训练定义一些操作。

```python
train_data_reader = lines_from_file(train_txt, repeat=True)

image_paths_placeholder = tf.placeholder(tf.string, shape=(None,
num_frames), name='image_paths')
labels_placeholder = tf.placeholder(tf.int64, shape=(None,),
name='labels')

train_input_queue =
data_flow_ops.RandomShuffleQueue(capacity=10000,
```

```
min_after_dequeue=batch_size,
dtypes= [tf.string, tf.int64],
shapes= [(num_frames,), ()])

train_enqueue_op =
train_input_queue.enqueue_many([image_paths_placeholder,
labels_placeholder])

frames_batch, labels_batch = input_pipeline(train_input_queue,
batch_size=batch_size, image_size=image_size)

with tf.variable_scope("models") as scope:
    logits, _ = nets.inference(frames_batch, is_training=True)

total_loss, cross_entropy_loss, reg_loss =
models.compute_loss(logits, labels_batch)
train_accuracy = models.compute_accuracy(logits, labels_batch)

global_step = tf.Variable(0, trainable=False)
learning_rate = models.get_learning_rate(global_step,
    initial_learning_rate, decay_steps, decay_rate)
train_op = models.train(total_loss, learning_rate, global_step)
```

在这段代码中，首先从文本文件中获得一个生成器对象，然后，为 image_paths 和 labels 创建两个占位符，它们将会被加入到队列 RandomShuffleQueue 中。在 datasets.py 中创建的 input_pipeline 函数将会接收 RandomShuffleQueue，并返回一批帧和标签。最后，创建计算损失值、准确率以及训练的其他相关操作。

为了记录训练过程，并在 TensorBoard 中对其进行可视化，我们将创建一些摘要。

```
tf.summary.scalar("learning_rate", learning_rate)
tf.summary.scalar("train/accuracy", train_accuracy)
tf.summary.scalar("train/total_loss", total_loss)
tf.summary.scalar("train/cross_entropy_loss", cross_entropy_loss)
tf.summary.scalar("train/regularization_loss", reg_loss)

summary_op = tf.summary.merge_all()

saver = tf.train.Saver(max_to_keep=max_checkpoints_to_keep)
time_stamp = datetime.now().strftime("single_%Y-%m-%d_%H-%M-%S")
checkpoints_dir = os.path.join(save_dir, time_stamp)
summary_dir = os.path.join(checkpoints_dir, "summaries")
train_writer = tf.summary.FileWriter(summary_dir, flush_secs=10)
```

```python
    if not os.path.exists(save_dir):
        os.mkdir(save_dir)
    if not os.path.exists(checkpoints_dir):
        os.mkdir(checkpoints_dir)
    if not os.path.exists(summary_dir):
        os.mkdir(summary_dir)
```

saver 和 train_writer 分别负责保存检查点和摘要。现在，创建会话并执行训练循环来完成训练过程。

```python
config = tf.ConfigProto()
config.gpu_options.allow_growth = True

with tf.Session(config=config) as sess:
    coords = tf.train.Coordinator()
    threads = tf.train.start_queue_runners(sess=sess, coord=coords)

    sess.run(tf.global_variables_initializer())

    num_batches = int(epoch_size / batch_size)

    for i_epoch in range(num_epochs):
        for i_batch in range(num_batches):
            # Prefetch some data into queue
            if i_batch % train_enqueue_steps == 0:
                num_samples = batch_size * (train_enqueue_steps + 1)

                image_paths, labels = \
sample_videos(train_data_reader, root_folder=train_folder,
num_samples=num_samples, num_frames=num_frames)
                print("\nEpoch {} Batch {} Enqueue {} \
videos".format(i_epoch, i_batch, num_samples))

                sess.run(train_enqueue_op, feed_dict={
                    image_paths_placeholder: image_paths,
                    labels_placeholder: labels
                })

            if (i_batch + 1) >= start_test_step and (i_batch + 1) % \
test_steps == 0:
                _, lr_val, loss_val, ce_loss_val, reg_loss_val, \
summary_val, global_step_val, train_acc_val = sess.run([
                    train_op, learning_rate, total_loss,
```

```
                cross_entropy_loss, reg_loss,
                        summary_op, global_step, train_accuracy
                    ])
                    train_writer.add_summary(summary_val,
global_step=global_step_val)
                    print("\nEpochs {}, Batch {} Step {}: Learning Rate
{} Loss {} CE Loss {} Reg Loss {} Train Accuracy {}".format(
                        i_epoch, i_batch, global_step_val, lr_val,
loss_val, ce_loss_val, reg_loss_val, train_acc_val
                    ))
                else:
                    _ = sess.run(train_op)
                    sys.stdout.write(".")
                    sys.stdout.flush()

            if (i_batch + 1) > 0 and (i_batch + 1) % save_steps ==
0:
                    saved_file = saver.save(sess,
os.path.join(checkpoints_dir, 'model.ckpt'),
                                            global_step=global_step)
                    print("Save steps: Save to file %s " % saved_file)

    coords.request_stop()
    coords.join(threads)
```

这段代码非常直观。我们将使用 `sample_videos` 函数获得图片路径和标签的列表。然后，调用 `train_enqueue_op` 操作将这些图片路径和标签添加到 RandomShuffleQueue 中。在此之后，就可以使用 `train_op` 来运行训练过程了，而且无须使用 `feed_dict` 机制。

现在，在 root 文件夹中调用下面的命令来运行训练过程。

export PYTHONPATH=.
python scripts/train.py

如果你的 GPU 内存大小不足以处理尺寸为 32 的批量数据，那么可能会遇到 OUT_OF_MEMORY 错误。在训练过程中，我们利用 `gpu_options.allow_growth` 创建了一个会话，你可以尝试改变 `batch_size` 来有效地使用 GPU 内存。

训练过程需要花费几小时才会收敛。我们会在 TensorBoard 中查看训练过程。

在用以保存检查点的目录中，运行以下命令。

tensorboard --logdir .

现在，打开 Web 浏览器并导航到 http://localhost:6006，结果如图 10-3 所示。

单 GPU 下的正则化损失和总损失如图 10-4 所示。

正如在这些图片中所看到的，数据训练的准确率需要大约 10 000 步才能达到 100%。在我的计算机上，这 10 000 步花费了 6 小时，在你的计算机上所花费的时间可能会不同。

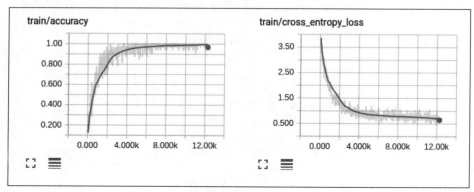

图 10-3

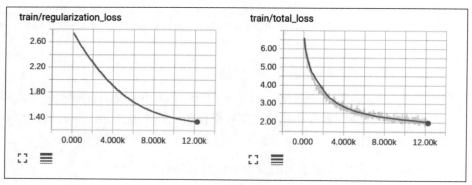

图 10-4

可以看到，训练的损失值在减小，如果训练时间更长，它可能会继续减小。然而，在 10 000 步之后，训练准确率几乎不再变化。

现在，我们使用多个 GPU 进行训练，并观察效果如何。

10.2.5　多 GPU 训练流程

我们的试验将使用自定义的机器而不是 Amazon EC2。然而，你可以在任何带有 GPU 的服务器上取得相同的结果。在本节中，我们将使用两个 Titan X GPU，每个 GPU 上的批量尺寸大小为 32 个视频。这样，每步可以计算 64 个视频，而不是单 GPU 配置下的 32 个视频。

现在，在 scripts 包中创建一个新的 Python 文件 train_multi.py。在该文件中添加以下代码来定义一些参数。

```
import tensorflow as tf
import os
import sys
from datetime import datetime
from tensorflow.python.ops import data_flow_ops

import nets
import models
from utils import lines_from_file
from datasets import sample_videos, input_pipeline

# Dataset
num_frames = 10
train_folder = "/home/aiteam/quan/datasets/ucf101/train/"
train_txt = "/home/aiteam/quan/datasets/ucf101/train.txt"

# Learning rate
initial_learning_rate = 0.001
decay_steps = 1000
decay_rate = 0.7

# Training
num_gpu = 2

image_size = 112
batch_size = 32 * num_gpu
num_epochs = 20
epoch_size = 28747

train_enqueue_steps = 50

save_steps = 200 # Number of steps to perform saving checkpoints
test_steps = 20 # Number of times to test for test accuracy
start_test_step = 50

max_checkpoints_to_keep = 2
save_dir = "/home/aiteam/quan/checkpoints/ucf101"
```

除了 batch_size，其他参数与前面 train.py 文件中的参数相同。在本试验中，我们将利用多个 GPU，使用数据并行策略来进行训练。我们将使用的批量尺寸大小为 64，而不是 32。将批量分成两部分，每部分都由一个 GPU 处理。然后，合并两个 GPU 的梯度以更新网络的权重和偏差。

接下来，使用与前面相同的操作，如下所示。

```
train_data_reader = lines_from_file(train_txt, repeat=True)

image_paths_placeholder = tf.placeholder(tf.string, shape=(None, num_frames), name='image_paths')
labels_placeholder = tf.placeholder(tf.int64, shape=(None,), name='labels')

train_input_queue = data_flow_ops.RandomShuffleQueue(capacity=10000, min_after_dequeue=batch_size,
dtypes= [tf.string, tf.int64],
shapes= [(num_frames,), ()])

train_enqueue_op = train_input_queue.enqueue_many([image_paths_placeholder, labels_placeholder])

frames_batch, labels_batch = input_pipeline(train_input_queue, batch_size=batch_size, image_size=image_size)

global_step = tf.Variable(0, trainable=False)
learning_rate = models.get_learning_rate(global_step, initial_learning_rate, decay_steps, decay_rate)
```

Now, instead of creating a training operation with 'models.train', we will create a optimizer and compute gradients in each GPU.

```
optimizer = tf.train.AdamOptimizer(learning_rate=learning_rate)

total_gradients = []

frames_batch_split = tf.split(frames_batch, num_gpu)
labels_batch_split = tf.split(labels_batch, num_gpu)
for i in range(num_gpu):
    with tf.device('/gpu:%d' % i):
        with tf.variable_scope(tf.get_variable_scope(), reuse=(i > 0)):
            logits_split, _ = nets.inference(frames_batch_split[i], is_training=True)
```

```
            labels_split = labels_batch_split[i]
            total_loss, cross_entropy_loss, reg_loss =
models.compute_loss(logits_split, labels_split)
            grads = optimizer.compute_gradients(total_loss)
            total_gradients.append(grads)
            tf.get_variable_scope().reuse_variables()

with tf.device('/cpu:0'):
    gradients = models.average_gradients(total_gradients)
    train_op = optimizer.apply_gradients(gradients, global_step)

    train_accuracy = models.compute_accuracy(logits_split,
labels_split)
```

程序将会在每个 GPU 上计算梯度，并将其添加到一个名为 total_gradients 的列表中，接着在 CPU 上使用 average_gradients 计算最终的梯度（稍后我们会创建 average_gradients）。然后，通过在优化器上调用 apply_gradients 来创建训练操作。

现在，将下面的函数添加到 root 文件夹下的 models.py 文件中，以计算 average_gradients。

```
def average_gradients(gradients):
    average_grads = []
    for grad_and_vars in zip(*gradients):
        grads = []
        for g, _ in grad_and_vars:
            grads.append(tf.expand_dims(g, 0))

        grad = tf.concat(grads, 0)
        grad = tf.reduce_mean(grad, 0)

        v = grad_and_vars[0][1]
        grad_and_var = (grad, v)
        average_grads.append(grad_and_var)
    return average_grads
```

然后，回到 train_multi.py 文件。同之前一样，我们将创建 saver 和 summary 来保存检查点和摘要。

```
tf.summary.scalar("learning_rate", learning_rate)
tf.summary.scalar("train/accuracy", train_accuracy)
tf.summary.scalar("train/total_loss", total_loss)
tf.summary.scalar("train/cross_entropy_loss", cross_entropy_loss)
tf.summary.scalar("train/regularization_loss", reg_loss)
```

```python
summary_op = tf.summary.merge_all()

saver = tf.train.Saver(max_to_keep=max_checkpoints_to_keep)
time_stamp = datetime.now().strftime("multi_%Y-%m-%d_%H-%M-%S")
checkpoints_dir = os.path.join(save_dir, time_stamp)
summary_dir = os.path.join(checkpoints_dir, "summaries")

train_writer = tf.summary.FileWriter(summary_dir, flush_secs=10)

if not os.path.exists(save_dir):
    os.mkdir(save_dir)
if not os.path.exists(checkpoints_dir):
    os.mkdir(checkpoints_dir)
if not os.path.exists(summary_dir):
    os.mkdir(summary_dir)
```

最后,添加训练循环来训练网络。

```python
config = tf.ConfigProto(allow_soft_placement=True)
config.gpu_options.allow_growth = True

sess = tf.Session(config=config)
coords = tf.train.Coordinator()
threads = tf.train.start_queue_runners(sess=sess, coord=coords)
sess.run(tf.global_variables_initializer())

num_batches = int(epoch_size / batch_size)

for i_epoch in range(num_epochs):
    for i_batch in range(num_batches):
        # Prefetch some data into queue
        if i_batch % train_enqueue_steps == 0:
            num_samples = batch_size * (train_enqueue_steps + 1)
            image_paths, labels = sample_videos(train_data_reader,
root_folder=train_folder,
num_samples=num_samples, num_frames=num_frames)
            print("\nEpoch {} Batch {} Enqueue {} videos".format(i_epoch, i_batch, num_samples))

            sess.run(train_enqueue_op, feed_dict={
                image_paths_placeholder: image_paths,
                labels_placeholder: labels
            })
```

```python
            if (i_batch + 1) >= start_test_step and (i_batch + 1) % test_steps == 0:
                _, lr_val, loss_val, ce_loss_val, reg_loss_val, summary_val, global_step_val, train_acc_val = sess.run([
                    train_op, learning_rate, total_loss, cross_entropy_loss, reg_loss,
                    summary_op, global_step, train_accuracy
                ])
                train_writer.add_summary(summary_val, global_step=global_step_val)

                print("\nEpochs {}, Batch {} Step {}: Learning Rate {} Loss {} CE Loss {} Reg Loss {} Train Accuracy {}".format(
                    i_epoch, i_batch, global_step_val, lr_val, loss_val,
                 ce_loss_val, reg_loss_val, train_acc_val
                ))
            else:
                _ = sess.run([train_op])
                sys.stdout.write(".")
                sys.stdout.flush()

            if (i_batch + 1) > 0 and (i_batch + 1) % save_steps == 0:
                saved_file = saver.save(sess,
                                        os.path.join(checkpoints_dir, 'model.ckpt'),
                                        global_step=global_step)
                print("Save steps: Save to file %s " % saved_file)

coords.request_stop()
coords.join(threads)
```

此处的训练循环与之前的类似，不过有一点不同：此处在会话配置中增加了 `allow_soft_placement=True` 选项，该选项将允许 TensorFlow 在必要的情况下改变变量的位置。

运行训练脚本。

`python scripts/train_multi.py`

经过几小时的训练后，查看 TensorBoard 来比较结果，如图 10-5 所示。

如你所见，在我们的计算机上运行大概 4 小时、约 6 000 步之后，在多 GPU 上的训练的准确率达到了 100%，这几乎减少了一半的训练时间。

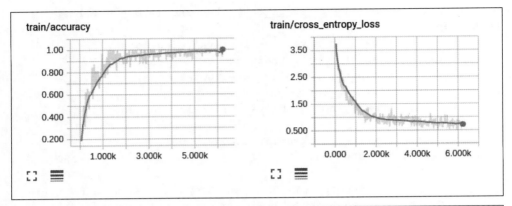

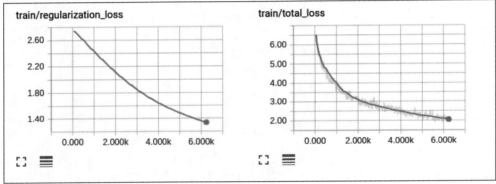

图 10-5

这两种训练策略的对比结果如图 10-6 所示。

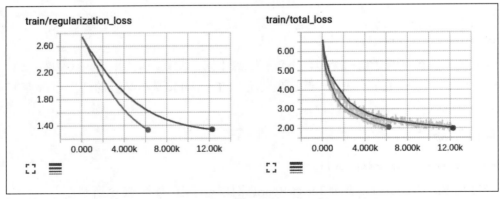

图 10-6

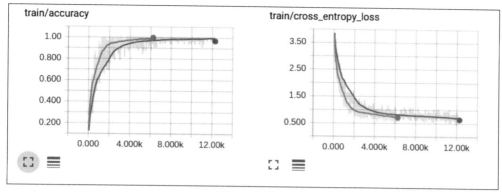

图 10-6（续）

在图 10-6 中，橙色的线是多 GPU 的训练结果，蓝色的线是单 GPU 的训练结果（彩图可从异步社区下载）。可以看到，多 GPU 设置能够取得更好的结果，但是差别不是很大。然而，利用更多的 GPU 可以获得更快的训练效果。在 Amazon EC2 的 P1 实例上，甚至存在 8 个和 16 个 GPU。如果在大型数据集上进行训练，例如 ActivityNet 或 Sports 1M，那么在多个 GPU 上的训练效果会更好，而单 GPU 则要花费很长时间才能收敛。

下一节，我们将快速浏览另一个亚马逊服务 Mechanical Turk。

10.3　Mechanical Turk 概览

Mechanical Turk 是一种服务，它允许我们创建和管理将由工作人员来完成的在线人工智能任务。工作人员在很多任务中能够比计算机做得更好。因此，可以利用这项服务来支持我们的机器学习系统。

下面是两个任务示例，可以使用它们来支持机器学习系统。

- **数据集标记**：通常你会有很多未标记的数据，此时可以使用 Mechanical Turk 来协助你为机器学习工作流建立一个一致的真实值数据。
- **生成数据集**：你可以请求工作人员创建大量的训练数据。例如，可以请求工作人员为一个自然语言系统创建文本翻译或聊天句子。另外，也可以让他们注释评论中的情绪。

除了标记，Mechanical Turk 还可以厘清混乱的数据集，以用于训练、数据分类和元数据标记，甚至可以使用该服务来判断系统输出。

10.4 总结

在本章中,我们学习了 Amazon EC2 服务,并了解了可以使用的服务器类型数量。然后,我们创建了一个神经网络,实现了在单个 GPU 上执行人体视频动作识别的任务。此外,还应用了数据并行策略来加快训练过程。最后,我们快速介绍了 Mechanical Turk 服务。希望读者能够利用这些服务,将机器学习系统提升到一个更高的水平。

第 11 章
更进一步——21 个课题

在本章中,我们将介绍 21 个可以通过深度学习和 TensorFlow 解决的现实课题。首先,我们会讨论一些公共的大型数据集和竞赛;然后,展示 GitHub 上一些出色的 TensorFlow 项目。另外,我们还将介绍一些用其他深度学习框架实现的有趣项目,从中你可以获得灵感并实现自己的 TensorFlow 解决方案。最后,我们会通过一种简单的技术将 Caffe 模型转换为 TensorFlow 模型,并介绍如何使用高级的 TensorFlow 库 TensorFlow-Slim。

在本章中,我们将探讨以下主题。

- ◆ 大规模的公共数据集和竞赛。
- ◆ 出色的 TensorFlow 项目。
- ◆ 由其他框架启发的一些深度学习项目。
- ◆ 将 Caffe 模型转换为 TensorFlow。
- ◆ 介绍 TensorFlow-Slim。

11.1 数据集和挑战赛

本节将展示一些流行的数据集和竞赛。

11.1.1 课题 1:ImageNet 数据集

ImageNet 是一项大型的视觉识别挑战赛,该挑战赛的数据集根据 WorkNet 结构组织,其中含有超过 1000 万张图片的 URL,并附带人工标注的标签来表示图片中的对象。另外,至少有 100 万张图片包含了边框。

自 2010 年起，ImageNet 挑战赛每年都会举行，并评估针对以下 3 个课题的算法。

- 针对 1 000 个类别的对象的定位。
- 针对 200 个标注好分类的对象的检测。
- 针对视频中 30 个标注好分类的对象的检测。2017 年挑战赛结果宣布了很多先进且有趣的算法。

11.1.2　课题 2：COCO 数据集

COCO 是一个由微软发布的用于图像识别、图像分割和图像标注的数据集。该数据集包含 80 个对象类别、超过 30 万张图片和 200 万个实例。另外，每年也会有针对检测、标注和关键点的挑战赛。

11.1.3　课题 3：Open Images 数据集

Open Images 是一个来自谷歌的新数据集，它包含覆盖 6 000 个类别的 900 多万张图像的 URL。每一张图像都经过了谷歌的视觉模型处理，并人工进行了验证。截止到 2017 年 7 月 20 日，已经有覆盖 600 个对象的超过 200 万个边界框被注解。

Open Images 比其他数据集覆盖了更多现实生活中的对象，这在开发实际应用时将会非常有用。

11.1.4　课题 4：YouTube-8M 数据集

YouTube-8M 是一个来自谷歌的大型视频数据集，其中包含 700 万个视频的 URL，4716 个类别、45 万小时的视频。谷歌也提供了预先计算的、先进的视听特征，这样用户就可以基于这些特征毫不费力地建立自己的模型。原始视频的训练可能需要花费数周时间，这在正常情况下是不合理的。该数据集的目标是实现视频理解、表征学习、含噪数据建模、迁移学习和视频的局部自适应学习等。

11.1.5　课题 5：AudioSet 数据集

AudioSet 是一个来自谷歌的大型音频事件数据集，其中包含 632 个音频类别，以及超过 210 万个人工标注的声音剪辑片段。音频类别包括人类声音和动物声音、乐器声音和普通的日常环境声音。利用该数据集，我们可以创建一个能够识别音频事件的系统，以用于音频理解、安全应用以及其他方面。

11.1.6 课题 6：LSUN 挑战赛

LSUN 挑战赛提供了一个大规模的场景理解数据集，主要覆盖以下 3 个问题：

- 场景分类；
- 街道图像分割任务；
- 显著性预测。

在场景分类问题中，算法期望的输出是图像展示的场景类别。在本书编写之际，该数据集一共有 10 种不同的类别，例如卧室、教室和餐馆等。在分割问题上，你可以试着去解决像素级的分割问题和特定实例的分割问题。显著性预测是预测一个人在场景图像中的位置。

11.1.7 课题 7：MegaFace 数据集

MegaFace 提供了一个大规模的人脸识别数据集。MegaFace 数据集分为 3 个部分：

- 训练集；
- 测试集；
- 干扰项。

训练集包含 470 万张（包含超过 672 057 个独特身份）的照片。测试集包含来自 FaceScrub 数据集和 FGNet 数据集中的图片。干扰项包含 100 万张（包含 690 572 个独特身份）的照片。目前，MegaFace 网站上有两个挑战。在挑战 1 中，你可以使用任何数据集进行训练，并使用干扰项来测试你的方法。你需要区分出一组已知的人，同时将这些干扰项分类为未知的人。在挑战 2 中，你将使用包含 672 057 个独特身份的训练集进行训练，并使用干扰项进行测试。在本书编写之际，MegaFace 是人脸识别领域最大的数据集之一。

11.1.8 课题 8：Data Science Bowl 2017 挑战赛

Data Science Bowl 2017 是一项针对肺癌检测的 100 万美元挑战的数据集。在该数据集中，你将会得到 1 000 多张高危病人的 CT 图像。这项挑战的目标是建立一个自动化的系统，该系统能够判断病人是否会在一年内被诊断为肺癌。这是一个非常有趣且十分重要的项目，在不久的将来，该项目可能会拯救成千上万的人。

11.1.9 课题 9：星际争霸游戏数据集

在本书编写之际，该数据集是最大的《星际争霸——母巢之战》录像数据集之一。该数据集大小为 365GB，包含 60 000 多个游戏录像、15.35 亿帧、4.96 亿个玩家动作。该数据集适合那些想要研究 AI 游戏的人。

11.2　TensorFlow 项目

在本节中，我们将介绍 GitHub 上几个开源的 TensorFlow 项目。建议你登录网站查看这些项目，并学习如何提高 TensorFlow 技能。

11.2.1　课题 10：人体姿态估计

该项目是 Deep Cut 和 ArtTrack 在人体姿态估计上的开源实现，其目标是解决检测和姿态估计任务。我们可以将该项目用到各种应用程序上，例如安防方面的人体检测或人体动作理解。另外，在深入研究人体形态估计在虚拟试衣和服装推荐的应用方面，该项目也提供了很好的起点。

11.2.2　课题 11：对象检测——YOLO

对象检测是计算机视觉中一个很有趣的问题。有很多方法可以解决这个问题，Joseph Redmon 等人开发的 YOLO 就是比较先进的技术之一。YOLO 利用深度神经网络提供了实时对象检测功能。YOLO 的第 2 版能够以高准确率实时识别多达 9 000 种不同对象。而最初的 YOLO 项目是在 darknet 框架下开发的。

TensorFlow 中有一个对 YOLO 很好的实现，叫作 darkflow。darkflow 仓库中甚至包含允许你导出模型并将模型在移动设备上运行以提供服务的功能。

11.2.3　课题 12：对象检测——Faster RCNN

Faster RCNN 是对象检测方面的另一个先进的方法。该方法能够获取准确度很高的结果，同时也给很多其他问题的解决方法提供了思路。Faster RCNN 的推理速度不像 YOLO 那么快。但是，如果你需要获取高准确度的检测结果，那么可以考虑使用 Faster RCNN。

11.2.4　课题 13：人体检测——Tensorbox

Tensorbox 是对 Russell Stewart 和 Mykhaylo Andriluka 提出的算法的一种 TensorFlow 实现。此方法的目标与前面方法的略有不同。Tensorbox 专注于解决群体人体检测问题。该方法使用一个递归的 LSTM 层来生成边框序列，并定义了一个新的损失函数来操作检测结果集。

11.2.5　课题 14：Magenta

Magenta 是谷歌大脑团队的一个项目，专注于使用深度学习生成音乐和其他艺术作品。这是一个非常活跃的代码库，里面包含很多有趣问题的实现，例如使图像风格化、生成旋律或者生成草图等。

11.2.6　课题 15：WaveNet

WaveNet 是一个来自谷歌 Deep Mind 的用于生成音频的神经网络架构。WaveNet 用于训练以生成原始的音频波形，并在文本-语音转换和音频生成方面取得了良好的成果。根据 Deep Mind 的消息，在美国英语和中国普通话的文本-语音转换问题上，WaveNet 将上述方法的性能与人类水平之间的差距缩小了 50% 以上。

11.2.7　课题 16：Deep Speech

Deep Speech 是一个开源的语音-文本转换引擎，它基于百度的一篇研究论文并通过 TensorFlow 实现。语音-文本转换是一个非常有趣的问题，而 Deep Speech 是解决该问题最先进的方法之一。通过 Mozilla 的 TensorFlow 实现，你甚至能够学习如何跨越多台机器使用 TensorFlow。然而，仍然存在一个问题，那就是个体研究人员无法访问大型的语音-文本转换数据集（与大公司的规模相同）。尽管我们可以使用 Deep Speech 或者自己实现一个转换引擎，但构建一个用于生产环境的良好模型仍然很困难。

11.3　有趣的项目

本节将展示在其他深度学习框架中实现的一些有趣的项目。这些项目在比较困难的问题上都取得了显著成果。如果你想挑战一下自己，那么可以尝试在 TensorFlow 中实现这些方法。

11.3.1　课题 17：交互式深度着色——iDeepColor

交互式深度着色（Interactive Deep Colorization，iDeepColor）是 Richard Zhang 和 Jun-Yan Zun 等人进行的一项研究，用于指导用户进行图像着色。在该系统中，针对图像中的某些点，用户向网络提供一些色彩提示，网络将会利用大规模数据中的语义信息来传播用户的输入值。此外，该研究已经可以在单个前进方向上进行实时着色。

11.3.2　课题 18：Tiny 人脸检测器

该项目是由 Peiyun Hu 和 Deva Ramanan 开发的一种人脸检测器，它专注于在图像中寻找小尺寸的人脸。大多数人脸检测器只关注图像中的大型对象，而这种小型人脸检测器方法则能够工作于人脸很小的环境。与之前方法相比，在 WIDER FACE 数据集上，它的误差是之前方法的 $\frac{1}{3}$。

11.3.3 课题 19：人体搜索

该项目是对 Tong Xiao 等人论文的实现，它关注人体检测和重识别问题，可用于视频监控。现有的人体重识别方法主要假设人物图像经过了剪裁和对齐处理。然而，在现实场景中，这些人体检测算法可能无法提取完美的人体裁剪区域，从而导致识别准确率降低。在该项目中，作者在 Faster RCNN 的启发下，以一种新颖的架构一并解决了检测和识别问题。当前该项目在深度学习框架 Caffe 中实现。

11.3.4 课题 20：人脸识别——MobileID

该项目提供了一个非常快速的人脸识别系统，该系统可以在帧率为每秒 250 帧的情况下取得很高的准确率。该模型是通过使用先进的人脸识别算法 DeepID 的输出而学习得到的。MobileID 模型的执行速度非常快，它可以用于处理能力和内存有限的场景中。

11.3.5 课题 21：问题回答——DrQA

DrQA 是 Facebook 的一个开放问答系统。DrQA 重点在于解决机器阅读的问题，它会试图理解维基百科文档，并回答用户的任何问题。当前该项目在 PyTorch 中实现。你会发现，在 TensorFlow 中实现我们自己的解决方案将会充满乐趣。

11.4 Caffe 转 TensorFlow

在本节中，我们将展示如何使用 Caffe Model Zoo 中的很多预训练模型，这里有很多针对不同任务使用各种架构的 Caffe 模型。将这些模型转换成 TensorFlow 后，就可以将转换后的模型当作架构的一部分，或者对我们的模型进行微调以用于不同的任务。相比于从头开始训练，使用这些预先训练过的模型作为初始权重是一种有效的训练方法。我们将展示如何使用一种将 Caffe 转为 Tensorflow 的方法，该方法来自 Saumitro Dasgupta。

然而，Caffe 和 TensorFlow 之间有很多不同之处。这种技术仅仅支持 Caffe 的部分层类型。不过，该项目的作者已经验证了一些 Caffe 架构（如 ResNet、VGG 和 GoogLeNet）可转化。

首先，需要使用 `git clone` 命令复制 caffe-tensorflow 仓库。

```
ubuntu@ubuntu-PC:~/github$ git clone
***://github.***/ethereon/caffe-tensorflow
Cloning into 'caffe-tensorflow'...
remote: Counting objects: 479, done.
remote: Total 479 (delta 0), reused 0 (delta 0), pack-reused 479
Receiving objects: 100% (510/510), 1.71 MiB | 380.00 KiB/s, done.
Resolving deltas: 100% (275/275), done.
Checking connectivity... done.
```

然后，将目录切换为 caffe-tensorflow，并运行 python 转换脚本以查看帮助信息。

```
cd caffe-tensorflow
python convert.py -h
The resulting console will look like this:
usage: convert.py [-h] [--caffemodel CAFFEMODEL]
                  [--data-output-path DATA_OUTPUT_PATH]
                  [--code-output-path CODE_OUTPUT_PATH] [-p PHASE]
                  def_path
positional arguments:
def_path                Model definition (.prototxt) path
optional arguments:
  -h, --help            show this help message and exit
  --caffemodel CAFFEMODEL
                        Model data (.caffemodel) path
  --data-output-path DATA_OUTPUT_PATH
                        Converted data output path
  --code-output-path CODE_OUTPUT_PATH
                        Save generated source to this path
  -p PHASE, --phase PHASE
                        The phase to convert: test (default) or train
```

根据这些帮助信息，你就可以知道 convert.py 脚本所需要的参数。总之，我们将使用 convert.py 脚本创建 TensorFlow 中的网络架构（使用标记 code-output-path 指定输出的路径），并转换预训练的权重（使用标记 data-output-path 指定输出的路径）。

在开始转换模型之前，我们需要从该项目的贡献者处获取一些信息。当前的 master 分支存在一些问题，即无法使用最新的 TensorFlow（本书编写之际为 1.3 版本）和 python-protobuf（本书编写之际为 3.4.0 版本）。我们将在 GitHub→ethereon/caffe-tensorflow/pull requests/105 或 133 处获取代码。

首先，从拉取请求 105 中获取代码。

```
ubuntu@ubuntu-PC:~/github$ git pull origin pull/105/head
remote: Counting objects: 33, done.
remote: Total 33 (delta 8), reused 8 (delta 8), pack-reused 25
Unpacking objects: 100% (33/33), done.
From ***//github.***/ethereon/caffe-tensorflow
 * branch              refs/pull/105/head -> FETCH_HEAD
Updating d870c51..ccd1a52
Fast-forward
 .gitignore                                        |  5 +++++
 convert.py                                        |  8 ++++++++
 examples/save_model/.gitignore                    | 11 ++++++++++
 examples/save_model/READMD.md                     | 17 +++++++++++++++++
 examples/save_model/__init__.py                   |  0
 examples/save_model/save_model.py                 | 51 ++++++++++++++++++++++++++++++++++++++++++++++
 kaffe/caffe/{caffepb.py => caffe_pb2.py}          |  0
 kaffe/caffe/resolver.py                           |  4 ++--
 kaffe/tensorflow/network.py                       |  8 ++++----
 9 files changed, 98 insertions(+), 6 deletions(-)
 create mode 100644 examples/save_model/.gitignore
 create mode 100644 examples/save_model/READMD.md
 create mode 100644 examples/save_model/__init__.py
 create mode 100755 examples/save_model/save_model.py
 rename kaffe/caffe/{caffepb.py => caffe_pb2.py} (100%)
```

然后，从拉取请求 133 中获取代码。

```
- git pull origin pull/133/head
remote: Counting objects: 31, done.
remote: Total 31 (delta 20), reused 20 (delta 20), pack-reused 11
Unpacking objects: 100% (31/31), done.
From ***//github.***/ethereon/caffe-tensorflow
 * branch              refs/pull/133/head -> FETCH_HEAD
Auto-merging kaffe/tensorflow/network.py
CONFLICT (content): Merge conflict in kaffe/tensorflow/network.py
Auto-merging .gitignore
CONFLICT (content): Merge conflict in .gitignore
Automatic merge failed; fix conflicts and then commit the result.
```

正如你所看到的，kaffe/tensorflow/network.py 文件存在一些冲突。接下来，我们将展示如何解决这些冲突，如下所示。

首先，解决第 137 行的冲突，如图 11-1 所示。

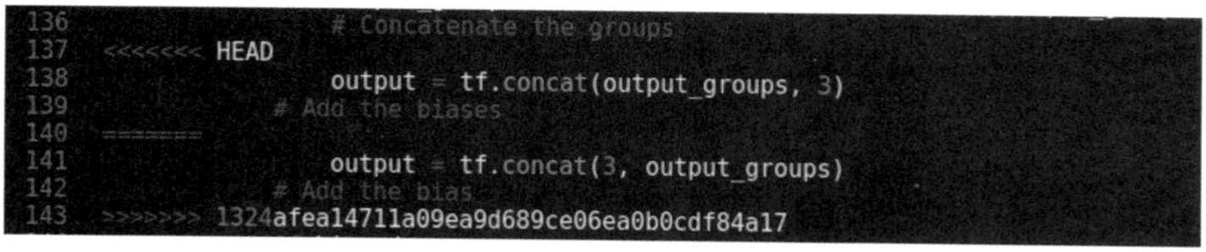

图 11-1

将第 137 行到 140 行之间的 HEAD 部分删除，最终的结果如图 11-2 所示。

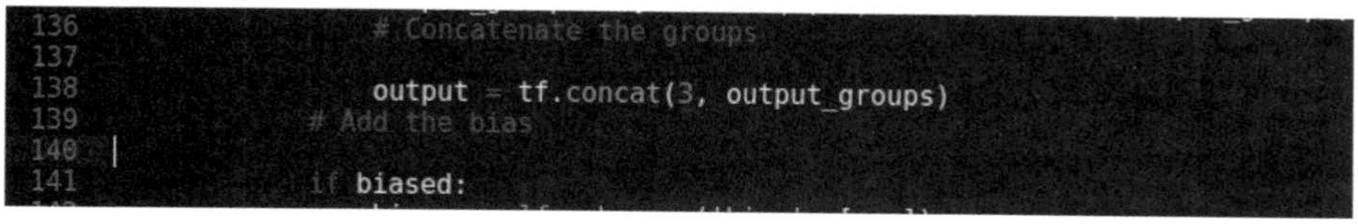

图 11-2

接下来，解决第 185 行的冲突，如图 11-3 所示。

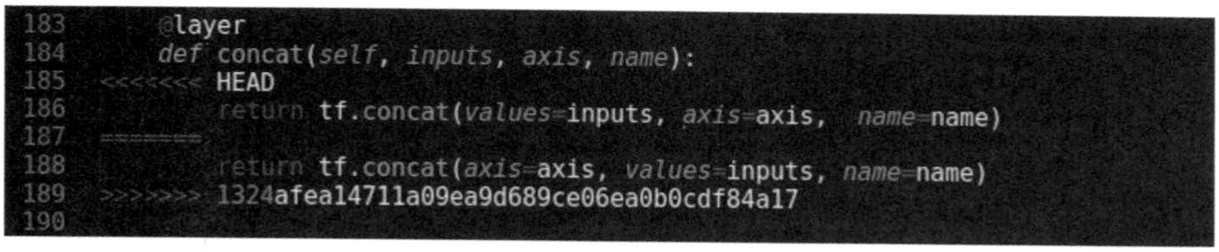

图 11-3

将第 185 行到 187 行之间的 HEAD 部分删除，最终的结果如图 11-4 所示。

图 11-4

在 caffe-tensorflow 目录中，存在一个名为 examples 的目录，该目录包含 MNIST 和 ImageNet 挑战的代码和数据。此处，我们将展示如何使用 MNIST 模型。ImageNet 挑战与 MNIST 模型没有太大区别。

首先，使用以下命令将 MNIST 架构从 Caffe 转换为 TensorFlow。

```
ubuntu@ubuntu-PC:~/github$ python ./convert.py
examples/mnist/lenet.prototxt --code-output-path=./mynet.py
    The result will look like this:
    ------------------------------------------------------------
        WARNING: PyCaffe not found!
        Falling back to a pure protocol buffer implementation.
        * Conversions will be drastically slower.
```

```
          * This backend is UNTESTED!
--------------------------------------------------------------
    Type                    Name
Param                   Output
--------------------------------------------------------------
    Input                   data                                  -
-       (64, 1, 28, 28)
    Convolution             conv1                                 -
-       (64, 20, 24, 24)
    Pooling                 pool1                                 -
-       (64, 20, 12, 12)
    Convolution             conv2                                 -
-       (64, 50, 8, 8)
    Pooling                 pool2                                 -
-       (64, 50, 4, 4)
    InnerProduct            ip1                                   -
-       (64, 500, 1, 1)
    InnerProduct            ip2                                   -
-       (64, 10, 1, 1)
    Softmax                 prob                                  -
-       (64, 10, 1, 1)
    Converting data...
    Saving source...
    Done.
```

然后，使用以下命令转换 MNIST 预训练的 Caffe 模型 examples/mnist/lenet_iter_10000.caffemodel。

```
ubuntu@ubuntu-PC:~/github$ python ./convert.py
examples/mnist/lenet.prototxt --caffemodel
examples/mnist/lenet_iter_10000.caffemodel --data-output-
path=./mynet.npy
```

结果如下所示。

```
          --------------------------------------------------------------
          WARNING: PyCaffe not found!
          Falling back to a pure protocol buffer implementation.
          * Conversions will be drastically slower.
          * This backend is UNTESTED!
          --------------------------------------------------------------
    Type                    Name
Param                   Output
```

```
    ---------------------------------------------------------------------
---------------------------
    Input                    data                                      -
-       (64, 1, 28, 28)
    Convolution              conv1
(20, 1, 5, 5)       (64, 20, 24, 24)
    Pooling                  pool1                                     -
-       (64, 20, 12, 12)
    Convolution              conv2
(50, 20, 5, 5)      (64, 50, 8, 8)
    Pooling                  pool2                                     -
-       (64, 50, 4, 4)
    InnerProduct             ip1
 (500, 800)         (64, 500, 1, 1)
    InnerProduct             ip2
(10, 500)           (64, 10, 1, 1)
    Softmax                  prob                                      -
-       (64, 10, 1, 1)
    Converting data...
    Saving data...
    Done.
```

如你所见，以上命令将会在当前目录中创建一个名为 mynet.py 的 Python 文件和一个名为 mynet.npy 的 NumPy 文件。另外，还需要将当前目录添加到 PYTHONPATH 中，以使其他代码能够正常导入 mynet.py。

```
ubuntu@ubuntu-PC:~/github$ export PYTHONPATH=$PYTHONPATH:.
ubuntu@ubuntu-PC:~/github$ python examples/mnist/finetune_mnist.py
....
('Iteration: ', 900, 0.0087626642, 1.0)
('Iteration: ', 910, 0.018495116, 1.0)
('Iteration: ', 920, 0.0029206357, 1.0)
('Iteration: ', 930, 0.0010091728, 1.0)
('Iteration: ', 940, 0.071255416, 1.0)
('Iteration: ', 950, 0.045163739, 1.0)
('Iteration: ', 960, 0.005758767, 1.0)
('Iteration: ', 970, 0.012100354, 1.0)
('Iteration: ', 980, 0.12018739, 1.0)
('Iteration: ', 990, 0.079262167, 1.0)
```

其中，每行的最后两个数字分别是微调过程的损失值和准确率。可以看到，利用 Caffe 模型中的预训练权重，微调过程的准确率可以轻易达到 100%。

查看 finetune_mnist.py 文件以分析预训练的权重是如何被使用的。

首先，使用以下代码导入 mynet。

```
from mynet import LeNet as MyNet
```

然后，为 `images` 和 `labels` 创建一些占位符，并使用 `ip2` 层计算损失值，如下所示。

```
images = tf.placeholder(tf.float32, [None, 28, 28, 1])
labels = tf.placeholder(tf.float32, [None, 10])
net = MyNet({'data': images})

ip2 = net.layers['ip2']
pred = net.layers['prob']

loss = tf.reduce_mean(tf.nn.softmax_cross_entropy_with_logits(logits=ip2, labels=labels), 0)
Finally, they load the numpy file into the graph, using the load method in the network class.
with tf.Session() as sess:
    # Load the data
    sess.run(tf.global_variables_initializer())
    net.load('mynet.npy', sess)
```

在此之后，微调过程就会独立于 Caffe 框架。

11.5 TensorFlow-Slim

TensorFlow-Slim 是 TensorFlow 中一个用于定义、训练和评估复杂模型的轻量级库。利用 TensorFlow-Slim，我们可以通过大量的高级网络层、变量和规则来更容易地构建、训练和评估模型。TensorFlow-Slim 库和 TensorFlow-Slim 预训练模型可在 GitHub 网站上的 TensorFlow 仓库中查看。

11.6 总结

在本章，我们介绍了很多有趣的挑战和问题，读者可以试着解决并学习它们，以提高自己的 TensorFlow 技能。在本章的末尾，我们还讲解了如何将 Caffe 模型转换为 TensorFlow，并介绍了高级的 TensorFlow 库 TensorFlow-Slim。

第 12 章 高级安装

因为深度学习涉及大量的矩阵乘法,所以当我们开始学习深度学习时,图形处理器(Graphic Processing Unit,GPU)就显得非常重要了。如果没有 GPU,那么试验过程可能要花费一天甚至更长时间。利用一个强大的 GPU,我们可以快速迭代深度学习网络和大型训练数据集,并能够在短时间内运行多个试验。利用 TensorFlow,我们可以轻易地在单个 GPU 甚至是多个 GPU 上工作。然而,一旦涉及 GPU,大多数机器学习平台的安装会变得十分复杂。

在本章,我们将讨论 GPU,并将重点放在 CUDA 设置和基于 GPU 的 TensorFlow 安装上。首先,安装 Nvidia 驱动程序、CUDA 工具箱和 cuDNN 库。然后,利用 pip 安装已开启 GPU 功能的 TensorFlow。最后,展示如何使用 Anaconda 进一步简化安装过程。

12.1 安装

在本章中,我们将在一台拥有 Nvidia Titan X GPU、系统为 Ubuntu 16.06 的计算机上工作。

 为了避免其他问题,建议使用 Ubuntu 14.04 或 16.06 系统。

GPU 的选择超出了本章的范围。然而,当与 CPU 比对时,为了充分利用 GPU 的优势,必须选择一个具有高存储容量的 Nvidia 设备。目前,TensorFlow 及大多数其他深度学习框架官方还不支持 AMD GPU。在本书编写之际,Windows 系统在 Python 3.5 或 Python 3.6

中可以使用支持 GPU 的 TensorFlow。然而，macOS 系统上 TensorFlow 从 1.2 版本开始放弃了对 GPU 的支持。如果使用的是 Windows 系统，建议按照官方教程来安装。

12.1.1 安装 Nvidia 驱动程序

在 Ubuntu 系统上安装 Nvidia 驱动程序有很多种方法。本节将展示最简单的一种方法，该方法使用专有的 GPU 驱动程序 PPA，它提供了稳定的、专有的 Nvidia 显卡驱动更新。

首先，打开终端并运行下面的命令，将 PPA 添加到 Ubuntu。

```
sudo add-apt-repository ppa:graphics-drivers/ppa
sudo apt update
```

现在，需要选择安装 Nvidia 驱动程序的某个版本。运行以下命令查看匹配计算机的最新版本。

```
sudo apt-cache search nvidia
```

上述命令执行的结果如图 12-1 所示。

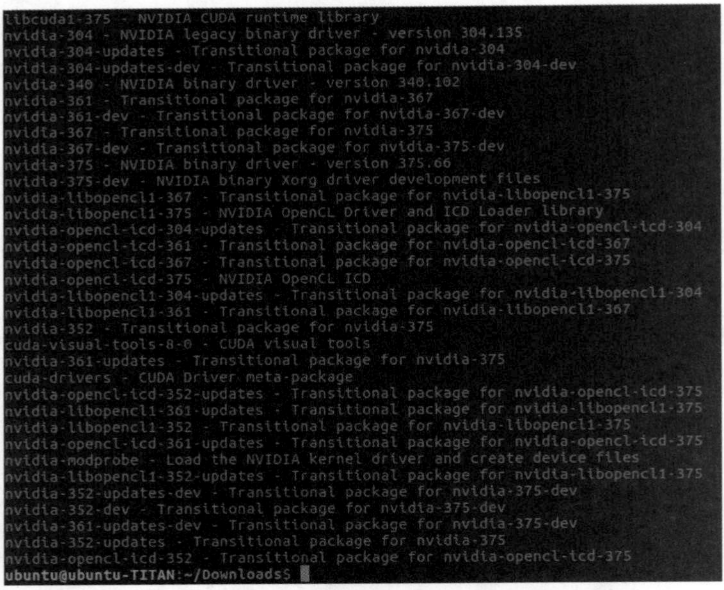

图 12-1

如你所见，本书所用计算机的最新的驱动程序版本是 375.66，它与 NVIDIA 二进制驱动程序文件名中的一致。现在，可以利用以下命令安装 Nvidia 驱动版本 375.66。

```
sudo apt-get install nvidia-375
```

上述命令执行的结果如图 12-2 所示。

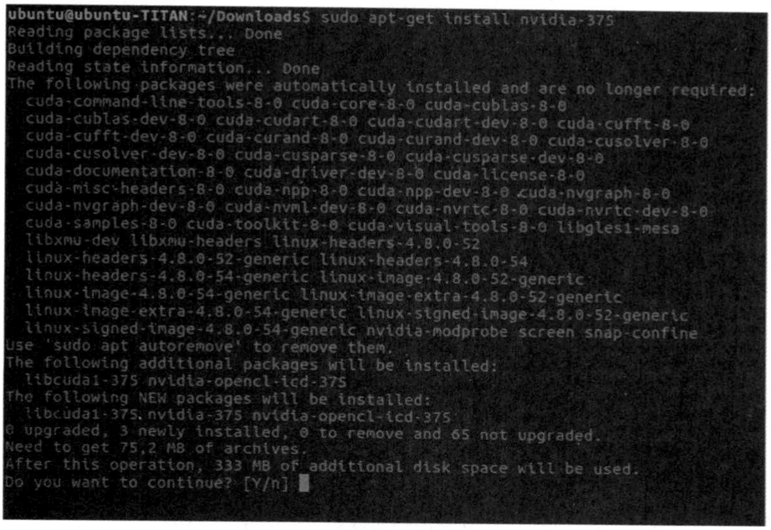

```
ubuntu@ubuntu-TITAN:~/Downloads$ sudo apt-get install nvidia-375
Reading package lists... Done
Building dependency tree
Reading state information... Done
The following packages were automatically installed and are no longer required:
  cuda-command-line-tools-8-0 cuda-core-8-0 cuda-cublas-8-0
  cuda-cublas-dev-8-0 cuda-cudart-8-0 cuda-cudart-dev-8-0 cuda-cufft-8-0
  cuda-cufft-dev-8-0 cuda-curand-8-0 cuda-curand-dev-8-0 cuda-cusolver-8-0
  cuda-cusolver-dev-8-0 cuda-cusparse-8-0 cuda-cusparse-dev-8-0
  cuda-documentation-8-0 cuda-driver-dev-8-0 cuda-license-8-0
  cuda-misc-headers-8-0 cuda-npp-8-0 cuda-npp-dev-8-0 cuda-nvgraph-8-0
  cuda-nvgraph-dev-8-0 cuda-nvml-dev-8-0 cuda-nvrtc-8-0 cuda-nvrtc-dev-8-0
  cuda-samples-8-0 cuda-toolkit-8-0 cuda-visual-tools-8-0 libgles1-mesa
  libxmu-dev libxmu-headers linux-headers-4.8.0-52
  linux-headers-4.8.0-52-generic linux-headers-4.8.0-54
  linux-headers-4.8.0-54-generic linux-image-4.8.0-52-generic
  linux-image-4.8.0-54-generic linux-image-extra-4.8.0-52-generic
  linux-image-extra-4.8.0-54-generic linux-signed-image-4.8.0-52-generic
  linux-signed-image-4.8.0-54-generic nvidia-modprobe screen snap-confine
Use 'sudo apt autoremove' to remove them.
The following additional packages will be installed:
  libcuda1-375 nvidia-opencl-icd-375
The following NEW packages will be installed:
  libcuda1-375 nvidia-375 nvidia-opencl-icd-375
0 upgraded, 3 newly installed, 0 to remove and 65 not upgraded.
Need to get 75.2 MB of archives.
After this operation, 333 MB of additional disk space will be used.
Do you want to continue? [Y/n]
```

图 12-2

安装结束后，应该会看到图 12-3 所示的画面。

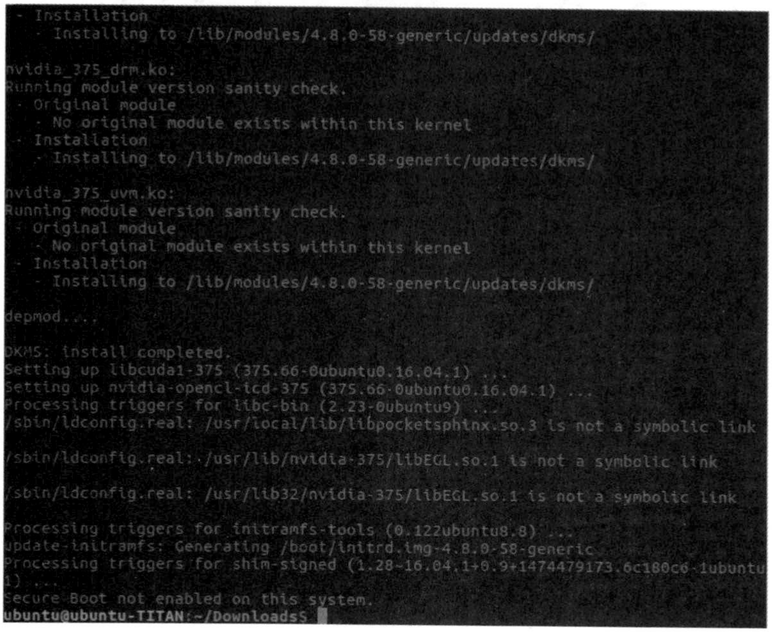

```
 - Installation
   - Installing to /lib/modules/4.8.0-58-generic/updates/dkms/

nvidia_375_drm.ko:
Running module version sanity check.
 - Original module
   - No original module exists within this kernel
 - Installation
   - Installing to /lib/modules/4.8.0-58-generic/updates/dkms/

nvidia_375_uvm.ko:
Running module version sanity check.
 - Original module
   - No original module exists within this kernel
 - Installation
   - Installing to /lib/modules/4.8.0-58-generic/updates/dkms/

depmod....

DKMS: install completed.
Setting up libcuda1-375 (375.66-0ubuntu0.16.04.1) ...
Setting up nvidia-opencl-icd-375 (375.66-0ubuntu0.16.04.1) ...
Processing triggers for libc-bin (2.23-0ubuntu9) ...
/sbin/ldconfig.real: /usr/local/lib/libpocketsphinx.so.3 is not a symbolic link

/sbin/ldconfig.real: /usr/lib/nvidia-375/libEGL.so.1 is not a symbolic link

/sbin/ldconfig.real: /usr/lib32/nvidia-375/libEGL.so.1 is not a symbolic link

Processing triggers for initramfs-tools (0.122ubuntu8.8) ...
update-initramfs: Generating /boot/initrd.img-4.8.0-58-generic
Processing triggers for shim-signed (1.28~16.04.1+0.9+1474479173.6c180c6-1ubuntu
1) ...
Secure Boot not enabled on this system.
ubuntu@ubuntu-TITAN:~/Downloads$
```

图 12-3

现在，我们将从 Nvidia 安装 CUDA 工具箱。

12.1.2 安装 CUDA 工具箱

首先,需要打开 Nvidia 网站下载 CUDA 工具箱,将看到图 12-4 所示的画面。

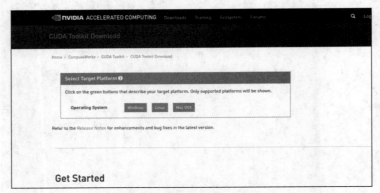

图 12-4

然后,选择"**Linux | x86_64 | Ubuntu | 16.04 | runfile(local)**",如图 12-5 所示。

图 12-5

接下来,单击 **Download [1.4 GB]** 按钮下载安装文件。安装文件的大小为 1.4 GB,下载将会需要一些时间。下载结束后，打开终端，将目录切换到包含该安装文件的路径中，并运行以下命令。

```
sudo sh cuda_8.0.61_375.26_linux.run
```

在命令行提示中，将看到"**End User License Agreement**"，如图 12-6 所示。

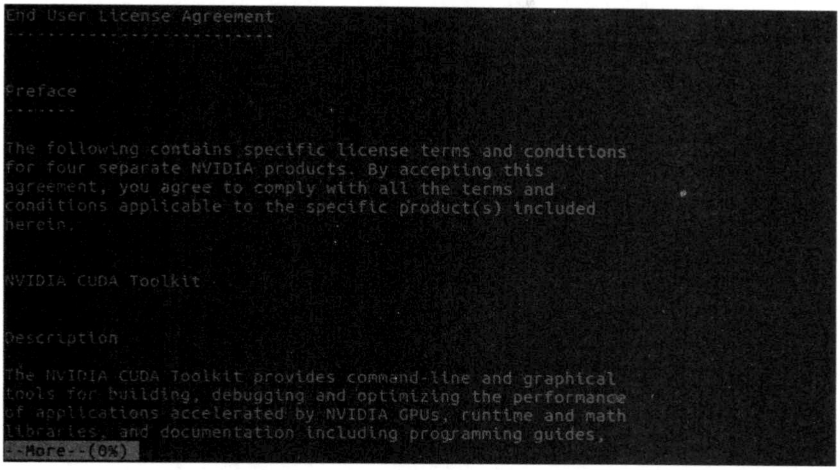

图 12-6

可以使用方向键来浏览协议。或者，也可以输入":q"，此时将会看到图 12-7 所示的画面。

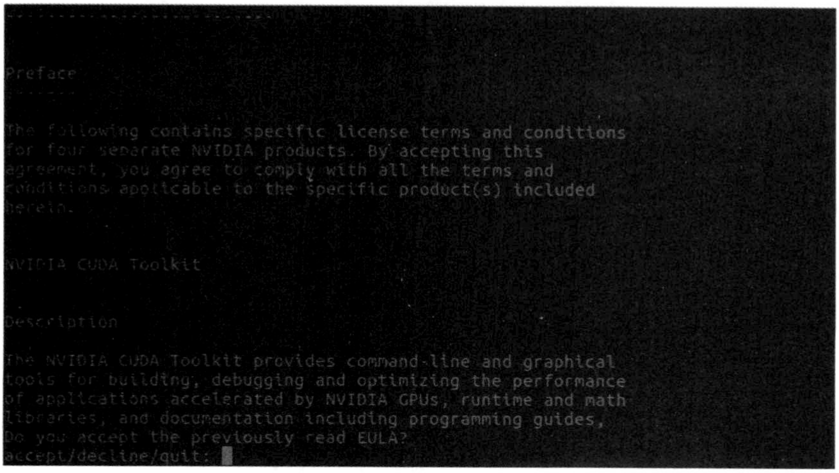

图 12-7

现在，可以输入"accept"接受该协议。在此之后，需要回答图 12-8 所示的一些问题。

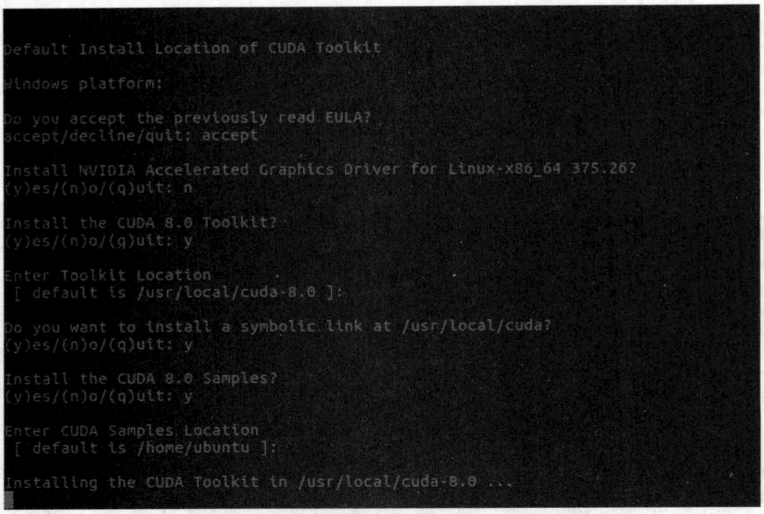

图 12-8

你可能会注意到,我们未在该提示中安装 Nvidia 驱动程序,这是因为在上一节中已经安装了最新的驱动程序。安装结束后,将看到图 12-9 所示的画面。

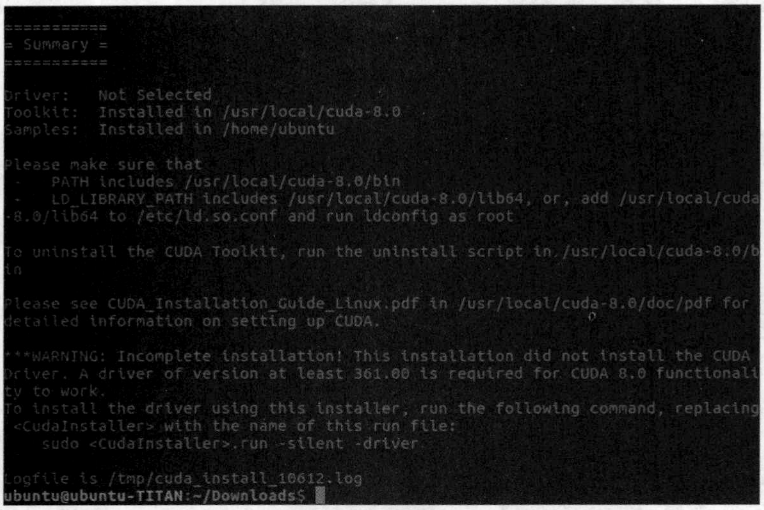

图 12-9

现在,打开~/.bashrc 文件,并在文件末尾添加下面内容。

export LD_LIBRARY_PATH=$LD_LIBRARY_PATH:/usr/local/cuda/lib64/

到此,我们已经成功地将 CUDA 工具箱安装到计算机中。此时,可以使用以下命令来查看显卡信息。

`nvidia-smi`

计算机上的结果如图 12-10 所示。

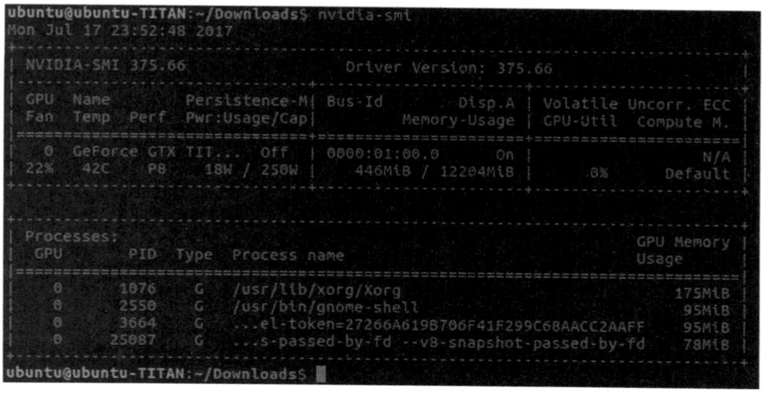

图 12-10

12.1.3 安装 cuDNN

为了使用支持 GPU 的 TensorFlow，需要从 Nvidia 安装另一个库 cuDNN。首先，需要浏览 Nvidia 网站，并下载 cuDNN 库。

你可能需要注册一个新的 Nvidia 账号。在登录 Nvidia 网站并打开 cuDNN 链接后，会看到图 12-11 所示的画面。

图 12-11

如你所见，cuDNN 存在几个版本，我们将使用 cuDNN v5.1 以适配 CUDA 8.0，这是 TensorFlow 需要的 cuDNN 版本。现在，可以通过链接"**cuDNN v5.1 Library for Linux**"下载该库，如图 12-12 所示。

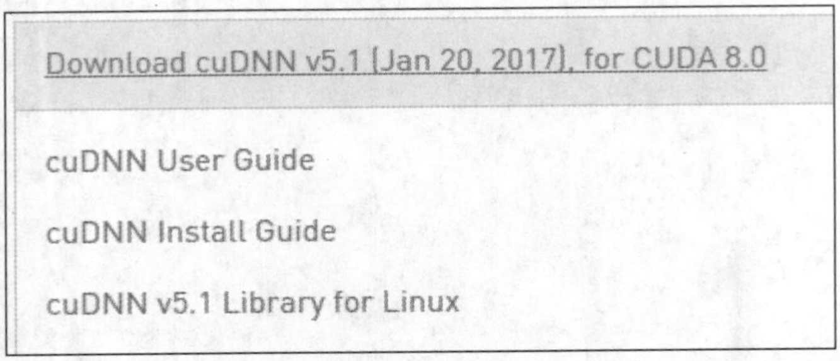

图 12-12

接着，在终端上使用以下命令将 cuDNN 安装到你的计算机上。

```
tar -xf cudnn-8.0-linux-x64-v5.1.tgz
cd cuda
sudo cp -P include/cudnn.h /usr/include/
sudo cp -P lib64/libcudnn* /usr/lib/x86_64-linux-gnu/
sudo chmod a+r /usr/lib/x86_64-linux-gnu/libcudnn*
```

运行结果如图 12-13 所示。

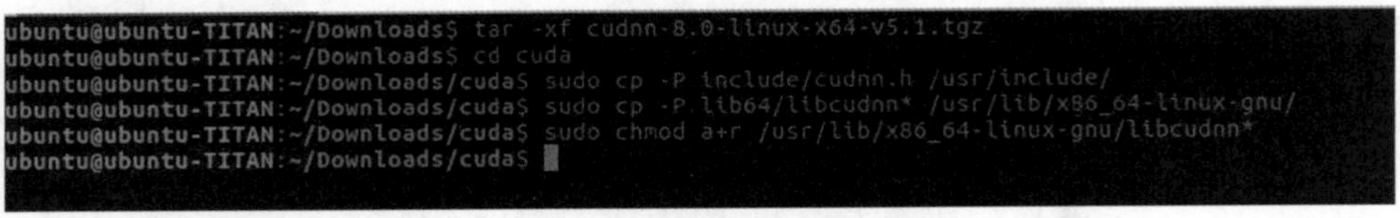

图 12-13

12.1.4　安装 TensorFlow

所有设置完成后，就可以使用 pip 工具轻松地安装支持 GPU 的 TensorFlow 了，如下所示。

```
sudo pip install tensorflow-gpu
```

命令执行结果如图 12-14 所示。

图 12-14

12.1.5 验证支持 GPU 的 TensorFlow

现在，可以在命令行中输入 `python`，并输入下面的 Python 命令，以检查 TensorFlow 是否能识别你的 GPU。

```
import tensorflow as tf
tf.Session()
```

结果应该如图 12-15 所示。

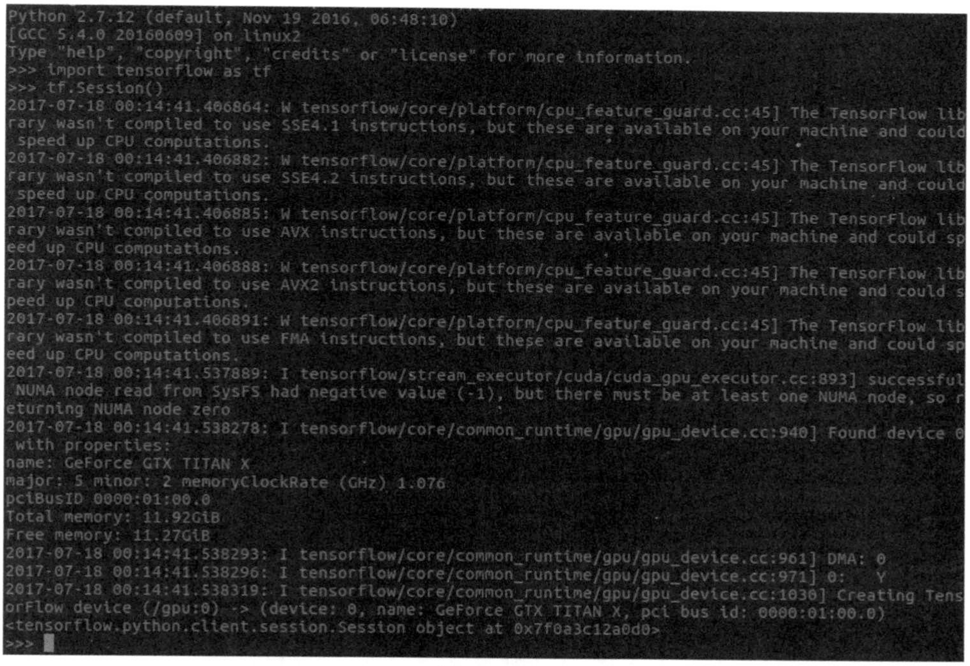

图 12-15

现在 TensorFlow 可以在 GPU 上工作了。TensorFlow 识别出我们的 GPU 为内存 11.92 GB 的 GeForce GTX TITAN X。下一节将展示利用多个版本的 TensorFlow 和工具库（例如 OpenCV）的方法。

12.2 利用 Anaconda 管理 TensorFlow

在工作过程中，我们会遇到在同一台计算机上需要多个版本的 TensorFlow 的情况，例如 TensorFlow 1.0 或 TensorFlow 1.2。另外，我们还可能需要在 Python 2.7 或 Python 3.0 下使用 TensorFlow。在之前的安装中，我们已经在 Python 系统中成功地安装了 TensorFlow。现在，我们将展示如何使用 Anaconda 在同一台计算机上管理多个工作环境。利用 Anaconda，甚至可以使用其他流行库的不同版本，例如 OpenCV、NumPy 和 scikit-learn 等。

首先，需要下载并安装 Miniconda。在我们的示例中，选择 Python 2.7 64 位的 bash 安装程序，这是因为我们希望使用 Python 2.7 作为默认的 Python。不过，后续可以使用 Python 2.7 或 Python 3.0 创建环境。我们需要运行以下命令来运行安装程序。

`bash Miniconda3-latest-Linux-x86_64.sh`

需要接受最终用户许可协议，如图 12-16 所示。

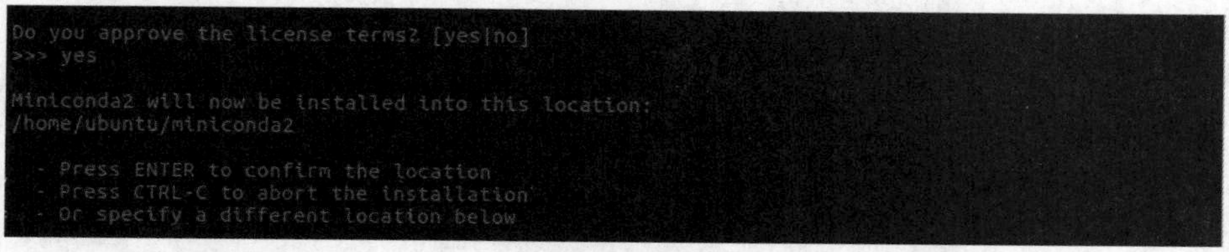

图 12-16

然后，继续安装。最终结果看起来应该如图 12-17 所示。

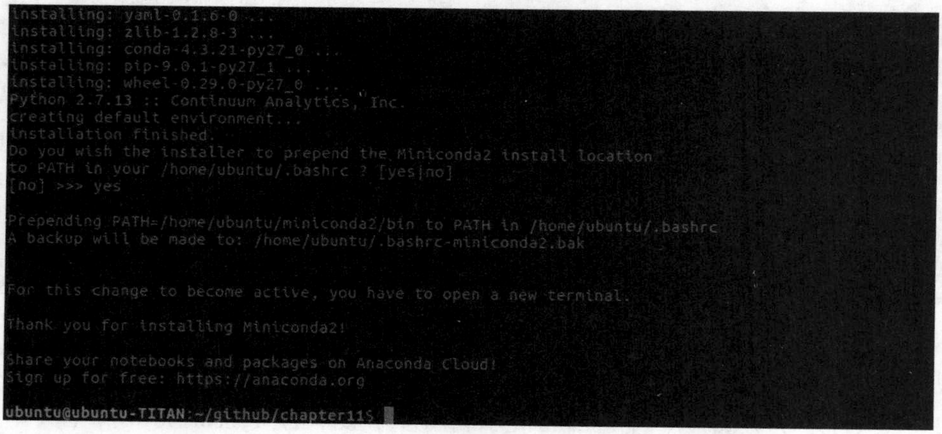

图 12-17

最后，需要引入 .bashrc 文件以准备并运行 Anaconda。

`source ~/.bashrc`

本章的源代码已经提供了一些环境配置，可以使用它们来创建期望的环境。

图 12-18 所示的是一个使用 Python 2.7、OpenCV 3 和支持 GPU 的 TensorFlow 1.2.1 的环境。该配置被命名为 env2.yml。

```
name: env2
channels:
    - https://conda.anaconda.org/menpo
    - conda-forge
dependencies:
    - python=2.7
    - opencv3
    - numpy
    - matplotlib
    - jupyter
    - pillow
    - scikit-learn
    - scikit-image
    - scipy
    - h5py
    - eventlet
    - flask-socketio
    - seaborn
    - pandas
    - ffmpeg
    - pyqt
    - pip:
        - moviepy
        - tensorflow-gpu
        - keras
        - prettytable
~
~
~
                                                  1,1           All
```

图 12-18

可以轻易地将 python=2.7 更改为 python=3，将 opencv3 更改为 opencv，以实现使用 python 3 和 OpenCV 2.4 的目的。

现在，运行以下命令来创建环境。

```
conda env create -f env2.yml
```

结果如图 12-19 所示。

```
Successfully installed backports.weakref-1.0rc1 funcsigs-1.0.2 imageio-2.1.2 keras-2.0.6 markdown-
2.6.8 mock-2.0.0 moviepy-0.2.3.2 pbr-3.1.1 prettytable-0.7.2 protobuf-3.3.0 tensorflow-gpu-1.2.1 t
heano-0.9.0 tqdm-4.11.2
#
# To activate this environment, use:
# > source activate env2
#
# To deactivate this environment, use:
# > source deactivate env2
#

ubuntu@ubuntu-TITAN:~/github/chapter11$
```

图 12-19

接下来，输入 `source activate env2` 来激活环境。

最后，同之前一样，需要验证 TensorFlow，如图 12-20 所示。

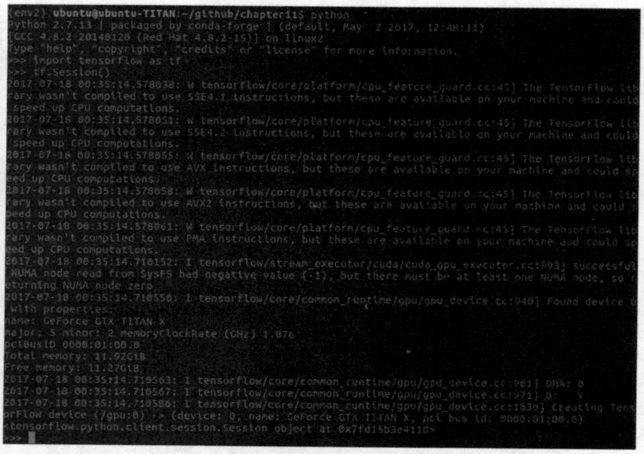

图 12-20

你可能会注意到图 12-20 左上角的"(env2)",它显示了当前环境的名称。第二行中显示的 Python 版本是 2.7.13,并由 conda-forge 封装。

现在,可以在工作流程中创建几个不同的环境以供使用。图 12-21 所示的是一个名为 env3 的环境示例,其中包含 Python 3 和 OpenCV 2.4。

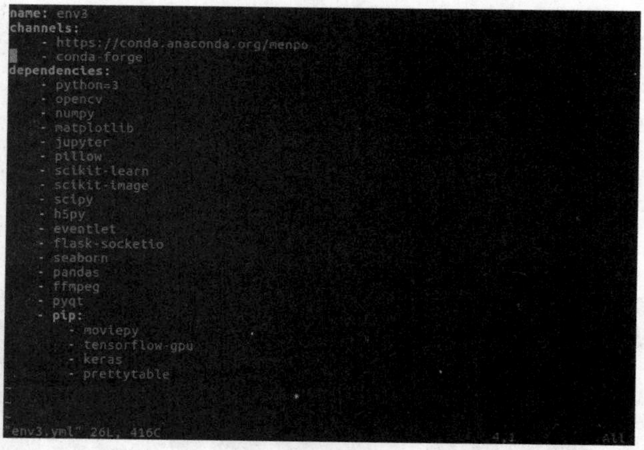

图 12-21

12.3 总结

在本章,我们讨论了在机器学习工作流中使用 GPU 的优势,尤其是在深度学习方面。然后,我们展示了 Nvidia 驱动、CUDA 工具箱、cuDNN 和支持 GPU 的 TensorFlow 的安装步骤。此外,我们还介绍了推荐使用多个版本的 TensorFlow 和其他库的工作流程。